Chaotic dynamics has been hailed as the third great scientific revolution in physics this century, comparable to relativity and quantum mechanics. In this book, Peter Smith takes a cool, critical look at such claims. He cuts through the hype and rhetoric by explaining some of the basic mathematical ideas in a clear and accessible way, and by carefully discussing the methodological issues which arise. In particular he explores the kinds of explanation of empirical phenomena which modern dynamics can deliver. *Explaining Chaos* will be complusory reading for philosophers of science and for anyone who has wondered about the conceptual foundations of chaos theory.

Peter Smith is Lecturer in Philosophy at the University of Cambridge. He has been editor of the journal *Analysis* since 1988, and is co-author (with O. R. Jones) of *The Philosophy of Mind* (Cambridge University Press, 1986) and co-editor (with Rosanna Keefe) of *Vagueness: A Reader* (MIT Press, 1997).

Explaining Chaos

Peter Smith

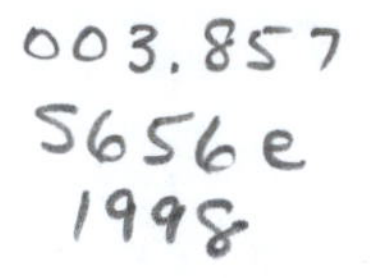

PUBLISHED BY THE PRESS SYNDICATE OF THE UNIVERSITY OF CAMBRIDGE
The Pitt Building, Trumpington Street, Cambridge, United Kingdom

CAMBRIDGE UNIVERSITY PRESS
The Edinburgh Building, Cambridge CB2 2RU, UK www.cup.cam.ac.uk
40 West 20th Street, New York, NY 10011–4211, USA www.cup.org
10 Stamford Road, Oakleigh, Melbourne 3166, Australia
Ruiz de Alarcón 13, 28014 Madrid, Spain

First published 1998
Reprinted 1999

Printed in the United Kingdom at the University Press, Cambridge

Typeset in Sabon 10pt and Myriad 9.5pt by the author

A catalogue record for this book is available from the British Library

Library of Congress Cataloguing in Publication data

ISBN 0 521 47171 0 hardback
ISBN 0 521 47747 6 paperback

Contents

Preface

I do not intend to avoid digressions and episodes; that is part of every conversation; indeed of life itself. Alexander Herzen (1968, 23)

'Chaos', in the sense that concerns us, is essentially a mathematical concept, and 'chaos theory' is a mathematical theory. Fully to grasp the concept or understand the theory means grappling with the relevant mathematics. A few basic ideas are explained in the course of this book, alongside the more conceptual or methodological discussions.

Relatively little background mathematics is presupposed in the main text. However, at various points it is illuminating (or perhaps simply fun) to go just a little further into the details in a way that may presuppose rather more – though familiarity with some calculus and the general idea of a differential equation, plus the ability to follow a moderately abstract mathematical argument, should largely suffice. These more taxing episodes are set in sans serif type and can be skimmed or skipped by the reader.

The same typographical device also serves to mark off other episodes that in a similar way pursue philosophical details rather further than some might want to follow. By picking and choosing among these passages, readers with various interests should be able to find a path through the book to suit.

To avoid further complicating the text, references to the mathematical and philosophical literature are kept to a minimum in the main body of the book; for a modest crop of additional references, see the final section, 'Notes'.

I have talked about these topics over the last few years at meetings associated with the British Society for the Philosophy of Science, in Bristol, London and Sheffield, and also at other seminars in Bradford, Keele, London, Nottingham, and Sheffield again. I am very grateful for all the comments that I have received from those various audiences.

But this book is especially associated with one particular place. It began life as a series of invited lectures in Cambridge in the Easter term, 1993; and more of the book was written for other events there – a Moral Sciences Club talk, an HPS seminar, a conference on the applications of mathematics, and finally another series of lectures in the Michaelmas term 1996. My warm thanks to many Cambridge friends – now colleagues – for all their help and criticism and advice.

Thanks also to Peter Dixon, Paul Glendinning, Adam Morton and to two anonymous readers for Cambridge University Press. I really ought to have completed the book sooner, as I was given British Academy funding for research leave to finish it in 1995, for which I am most grateful.

Not least, special thanks to Patsy Wilson-Smith, who had to live with chaos for too long.

Many sections of this book were sketched out during various stays in Trinity College. When I was a maths undergraduate there, two elderly inhabitants were pointed out (with a mixture of awe and irreverence, that being the sixties) – the philosopher C. D. Broad, and the mathematician A. S. Besicovitch. I wish I'd known something then about Besicovitch's work on what we now think of as fractal monsters. And I wish that I'd got to know Broad's work earlier too; his attitude to science, informed but cautious, admiring but refusing to be dazzled, still seems among the very best models for philosophers. I rather hope that their shades will look with benevolence on this book.

1

Chaos introduced

1.1 'Chaos theory': the very name suggests a paradox. For chaos, in the ordinary sense, is precisely the absence of order; and how can we hope to impose theoretically disciplined order on the essentially disordered?

It is far too late to change the name. So let's make it clear at the very outset that 'chaos' here must be taken as a term of art, stripped of most of its ordinary connotations: and 'chaos theory' is just the popular label for a body of theory about certain mathematical models and their applications. A first introductory task, then, is to say something about the kinds of mathematical models that are in question.

One mark of 'chaos' is *sensitive dependence on initial conditions*: that is to say, a chaotic system starting off from two very similar initial states can develop in radically divergent ways. Such sensitive dependence is often referred to as 'the Butterfly Effect'. A small blue butterfly, let's suppose, sits on a cherry tree in a remote province of China. As is the way of butterflies, while it sits it occasionally opens and closes its wings. It could have opened its wings twice just now; but in fact it moved them only once. And – because the weather system exhibits sensitive dependence – the minuscule difference in the resulting eddies of air around the butterfly eventually makes the difference between whether, two months later, a hurricane sweeps across southern England or harmlessly dies out over the Atlantic. Or so the story goes.

But it is no news that minuscule changes can have huge effects. I take aim at the President; the tiniest movement of my finger on the hair-trigger can change world history (and after the first few post-assassination hours, all quite unpredictably). If talk of a chaotic Butterfly Effect were merely more talk of the way that very small changes in initial conditions can get greatly amplified by later events in ways that prevent useful prediction, then we would be in entirely familiar territory. So a second introductory task is to explain where recent chaos theory gives an intriguing new twist to the old thought that micro-changes can produce unpredictable macro-effects.

1.2 Suppose we have a real-world phenomenon whose state at a particular time can be characterized by the values of the n variables $x_1, x_2, \ldots, x_n$ (so the x_i might represent the angular position and velocity of a swinging pendulum bob; or maybe they indicate the relative concentrations of certain chemicals in a mixture, or the velocity and temperature gradients in a convecting fluid, or ...). If we choose the right quantities to represent by these *state variables*, then we may be able to specify the dynamics of the phenomenon – the way that the phenomenon evolves over time – by giving the rate of change of each variable as some function of the x_i. In other words, we may be able to describe the dynamics by means of a set of n linked differential equations in the canonical form

$$\text{(D)} \quad dx_i/dt = F_i(x_1, x_2, \ldots, x_n) \qquad i = 1, \ldots, n.$$

Sometimes the F_i will be functions of time t as well: but we'll henceforth largely ignore that complication. More importantly for us, the F_i will typically be functions not only of the state variables x_i but also of some further *parameters*. These represent other quantities which are held constant while the x_i vary, but whose values can affect how the x_i evolve.

That is very schematic. But for a simple textbook example, take the case where we are modelling the behaviour of a freely swinging pendulum moving in a plane. Suppose θ represents the angular displacement of the bob from the vertical. Then, idealizing by ignoring the damping from air resistance etc., we have

$$d^2\theta/dt^2 = -(g/l)\sin\theta,$$

where l indicates the length of the pendulum, and g gives the gravitational force. Hence, putting $x_1 = d\theta/dt$ and $x_2 = \theta$, we have

$$\begin{aligned} dx_1/dt &= -(g/l)\sin x_2 \\ dx_2/dt &= x_1, \end{aligned}$$

a pair of equations now in the standard form (D). Here l and g are assumed to remain fixed while the pendulum is swinging, so we don't need equations for *their* rates of change: however, we may of course be very interested in the way that changes in the settings of these two parameters affect the way the pendulum swings.

If the F_i in the set of equations (D) satisfy some relatively mild constraints, then there will be a unique solution to the equations for given initial conditions. In other words, a particular setting of the parameters and the initial values of the x_i at time $t = t_0$ will fix a unique set of values for the x_i at least for some interval of time around

t_0 (and perhaps for *all* times). In the general case we will not be able to write down an explicit solution – we won't be able to specify the value of each of the x_i in terms of e.g. nice polynomial or trigonometrical functions of time. So to *use* the equations we will have to resort to numerical integration by computer. But the point of principle remains: the set of equations determines a unique evolution of the state variables over some stretch of time. Hence the equations describe a mathematical model which is, in a straightforward sense, deterministic.

It is extraordinarily helpful to look at things geometrically. So imagine treating the values of the n state-variables x_i as giving the coordinates of a point in an abstract n-dimensional space (a so-called state space or *phase space*). A point x in this phase space with the coordinates $x_1, x_2, \ldots, x_n$ will then represent a particular instantaneous state of our dynamical system. And given a point $x(0)$ representing the state at some initial time $t = t_0$, the dynamical equations (with fixed parameters) will, in the deterministic case, fix a unique trajectory or path traced out in phase space by the point $x(t)$ representing the state at later times t.

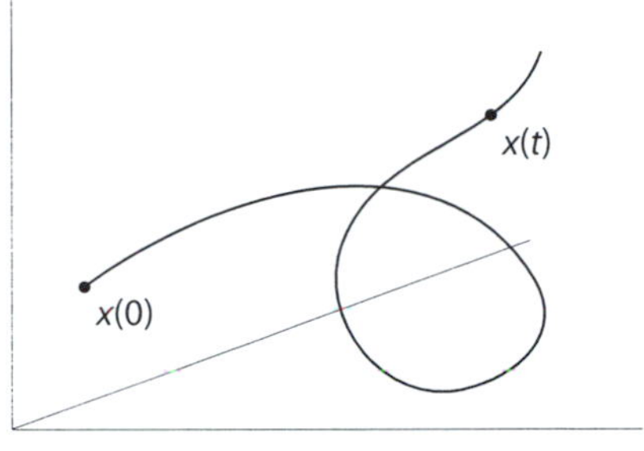

Figure 1.1 A phase space trajectory

A rich body of mathematics studies the behaviour of dynamical models given by sets of interlocking differential equations like (D). As we noted, there are usually no neat explicit solutions: but in fact we are often less concerned to find an explicit solution for a given set of initial conditions than to derive some general results about solutions. For example, we may be interested in the 'stability' of solutions. Does much the same happen if we run the model again from nearby starting points? Or to put it geometrically: do phase space trajectories which start close together tend to stay close together? And what happens if we run the model again with slightly different settings for the parameters in the equations: does the model behave in much the same way, or does the particular pattern of trajectories depend critically on the parameter values? What happens if we now add some random 'noise' to the equations: does the resulting phase space trajectory from a given starting point stay close to the original noiseless trajectory, or is a little noise sufficient to knock a trajectory far off course? Again, what happens to the behaviour in the model in the long term: does it (for example) tend to settle down into some regular, more or less periodic pattern – so that phase space trajectories eventually keep winding back on themselves, in the limit retracing their paths?

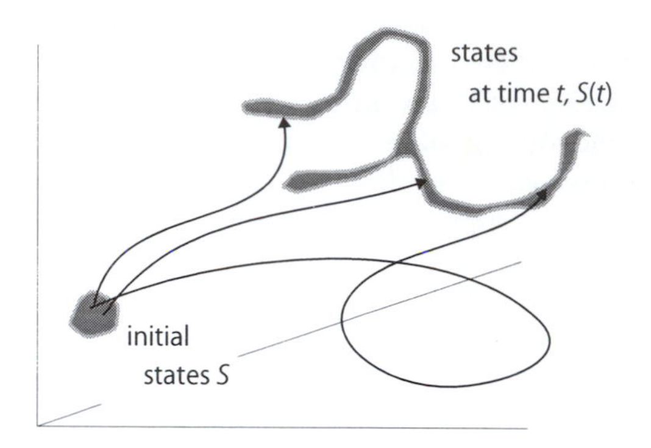

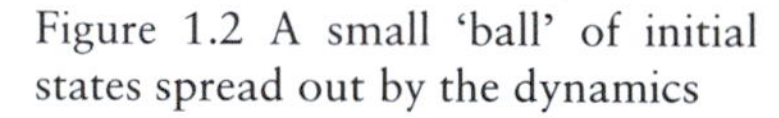
Figure 1.2 A small 'ball' of initial states spread out by the dynamics

We can often make headway on this sort of structural question even when exact solutions to our equations are elusive. And such structural issues are of central importance. Take just the first 'stability' question. If we are to apply a mathematical model to predict the evolution of some real-world dynamical phenomenon, then we must start by fixing the initial conditions to feed into the model. But we can only know the actual initial conditions within some margin of error. If we input a small error in representing the initial real-world state, then the dynamical equations will output a correspondingly erroneous prediction about where the system ends up at later time t (and the predictive error may very well grow over time). To put it geometrically, we can only pin down the point representing the initial state of the dynamical system to within some small fuzzy-boundaried 'ball' of phase space; and our dynamical equations will then map that fuzzy initial region of phase space onto a possibly much more spread-out region which will contain the point representing the later state at t (Figure 1.2). In order to make sensible use of a model predictively, then, we need to know something about just how much more spread out that later region is: we need to know, in other words, how quickly the dynamical model propagates initial errors. The interesting chaotic cases – as we will soon see – are cases where errors propagate exponentially fast.

Dynamical systems and dynamical equations

There are at least three distinct things that can be meant by talk of a 'dynamical system': a *real-world system which changes over time* (e.g. the planets in motion, a convecting fluid, a continuous chemical reaction, etc.), a *system of dynamical equations* which perhaps aims to capture the time-evolution of various quantities in some real-world system (though such equations may be studied simply for their mathematical interest), and an *abstract mathematical structure* (such as a 'flow' of trajectories through phase space) which is characterized by a set of dynamical equations. This multiple use is mostly harmless, but we should try to be more careful here. So from now on, let's reserve the term 'dynamical system' for when we are referring to some real-world dynamical phenomena. We will call systems of dynamical equations exactly that. And we will often refer to the abstract mathematical structures characterized by a set of equations as a mathematical *model*. (This last terminological choice reflects a doctrine to be explored later – namely that, in dynamical modelling, we are primarily comparing an abstract structure with some structure to be discerned in the target real-world dynamical system. So the model that gets applied *is* the structure described by the dynamical equations.)

Systems of dynamical equations can come in a variety of forms. They can be sets of one or more interlocking ordinary differential equations – of form (D), perhaps – or sets of partial differential equations. Or they can be sets of discrete difference equations which e.g. determine $x(n+1)$, the value of x at the $(n+1)$-th tick of the clock, in terms of $x(n)$ and perhaps $x(n-1)$ etc. Difference equations will feature in important ways later (§6.1). But for the moment, we will continue to consider only systems of ordinary differential equations.

As we saw in the case of the equation for an idealized pendulum, even if a system of equations is not initially presented in the form (D), we can usually knock it into shape by a judicious change of variables. For another quick example, take the superficially tangled equation

(U) $$2x'''x'^2 - 5x'''x''x' + 3x''^3 = 0,$$

where the prime indicates differentiation (so x' is just dx/dt etc.). By the simple dodge of putting $x = x_1, x' = x_2, x'' = x_3$, we find that (U) is equivalent to the linked equations

(U′) $$\begin{aligned} dx_1/dt &= x_2 \\ dx_2/dt &= x_3 \\ dx_3/dt &= 3x_3^2/x_2(5x_3 - 2x_2) \end{aligned}$$

and we are back to a system in canonical form. This particular system of equations, by the way, has a quite bizarre property (see Jackson 1990, vol. 1, 33–34). Take *any* continuous function $\phi(t)$: then (U′) has a (continuous, differentiable) solution $x(t)$ which close-tracks $\phi(t)$ – i.e. $|\phi(t) - x(t)| < \epsilon$ for arbitrarily small ϵ – for arbitrarily long. So (U′) – or equivalently (U) – is a kind of universal system guaranteed to have a solution to fit any empirical curve $\phi(t)$ as well as you could want. Such promiscuity of course makes a system like (U′) utterly useless in practice; it is mentioned here as a salutory illustration of the point that the behaviour of solutions of apparently simple equations in form (D) can be startlingly complex.

Still, we might have expected that, so long as the right hand side of a set of equations in form (D) is well-defined (so long as there is e.g. no division by zero), the equations will at least suffice to determine unique solutions for given initial conditions. If the initial values of $x_1, x_2, \ldots, x_n$ are fixed and the rates at which these variables change are a determinate function of their values, won't the evolving values of the variables be fixed thereby?

Not necessarily. Consider the one-variable equation

$$\text{(V)} \qquad dx/dt = s(x)$$

where $s(x) = 0$ when $x < 0$, and $s(x) = \sqrt{x}$ when $x \geqslant 0$. And suppose the initial condition is that $x = 0$ at time $t = 0$. This little system can 'burst into life' at any time $\tau \geqslant 0$: i.e. the function

$$\begin{aligned} x &= 0 && \text{for } t < \tau \\ x &= (t - \tau)^2/4 && \text{for } t \geqslant \tau \end{aligned}$$

is a solution of (V) for *any* positive kick-off time τ. So there is no unique solution, and determinism fails.

(This observation incidentally refutes over-casual assumptions to the effect that the Newtonian mechanics of a point particle governed by a reasonably nice-looking force function is always a deterministic theory. Imagine a system which is governed by the following Newtonian equation with a velocity dependent force:

$$m\frac{d^2y}{dt^2} = A\sqrt{\frac{dy}{dt}}, \qquad y \geqslant 0$$

Then, putting $x_1 = y$ and $x_2 = (m^2/A^2)dy/dt$, we derive a pair of equations of form (D)

$$\begin{aligned} dx_1/dt &= (A^2/m^2)x_2 \\ dx_2/dt &= \sqrt{x_2} && \text{for } x_2 \geqslant 0 \end{aligned}$$

which, as we have in effect just seen, will not determine a unique solution for the initial condition $x_1 = x_2 = 0$.)

Still, a set of first-order equations in canonical form *will* have a unique solution for a given initial condition so long as one or another of various pretty mild constraints are satisfied. For example, if the F_i are not only continuous but *continuously differentiable* in the neighbourhood of x^*, that will ensure that there is – at least locally – a unique solution with the value x^* at a given time t^*, and that this solution is defined for some period both before and after t^*. (This continuous differentiability constraint is not met in the case of (V) since the differential $ds(x)/dx$ is discontinuous at $x = 0$.)

Why qualify, and only say that there is a unique solution 'at least locally'? Well, consider the simple equation

$$dx/dt = x^2$$

and suppose that $x = a$ (where $a > 0$) at time $t = 0$. The function x^2 is continuously differentiable around $x = a$, so there is a unique solution with that initial condition, namely $x = a/(1 - at)$. This behaves nicely in the locality of $t = 0$; however, it blows up to infinity as $t \to 1/a$. A solution for a given initial condition may thus be unique but short-lived.

However, having noted for the record some of the things that can go wrong with even the simplest systems of equations in canonical form, we will concentrate henceforth on cases where things go right – i.e. where there *is* a unique solution for any given initial condition (and indeed we will typically be concerned with cases where a solution exists for all finite times). Or to put it geometrically: we will be interested in cases where the equations fix a unique phase space trajectory passing once through any given point, and where trajectories don't shoot off to infinity in finite time.

Finally, let's quickly note a result that will be important later (§4.3), namely that the same condition about continuous differentiability which ensures unique solutions also ensures that the solutions are *continuous in the initial data*. To explain: take the trajectory leading from $x(0)$ to $x(t)$: and assume that if the initial point $x(0)$ is nudged to some neighbouring point $y(0)$ then the point on the trajectory at time t gets moved to $y(t)$. The continuity result tells us that the distance $|x(t) - y(t)|$ can be made arbitrarily small by making $|x(0) - y(0)|$ small enough. Hence a smoothly continuous adjustment of the initial point doesn't produce a discontinuous jump in the state at time t: moving the start of a trajectory is like shaking the end of an elastic rope – the old trajectory will deform continuously into the new one. Of course, the fact that nudging the starting point $x(0)$ can not produce a *discontinuous* jump in $x(t)$ is quite consistent with small initial nudges producing extremely *large* changes in $x(t)$. In particular, the continuity result is still compatible with an error in setting $x(0)$ producing an error in the corresponding $x(t)$ that grows exponentially with time.

1.3 Suppose then that we are dealing with a set of equations which are in the canonical form (D) and which are everywhere deterministic; that is to say, for any starting set of values x^* there is a *unique* solution defined for some period around the time t such that $x(t) = x^*$. It is immediate that distinct phase space trajectories cannot cross or merge, nor can a single trajectory intersect itself. For if there were a crossing point or a merge point at x^* then the equations would have to allow more than one path through x^*, contradicting the deterministic assumption of local uniqueness.

The 'no crossing' rule puts constraints on what can happen in this sort of deterministic model. Consider first the 'one-dimensional' case where there is an equation in just one variable and the phase space is the real line R. How might the value of the single state variable x evolve over time? One possibility is that nothing happens – i.e. x always has the same fixed value. Another possibility is that the absolute value of x gets ever larger, and the trajectory wanders off towards infinity as $t \to \infty$. A third possibility is that a trajectory gets attracted to some fixed point a, so $x(t) \to a$ as $t \to \infty$ (e.g. the simple system $dx/dt = -x$ has the solution $x(t) = x(0)e^{-t}$, so $x(t) \to 0$ as t increases, whatever the initial $x(0)$). And because of the 'no-crossing' rule, that exhausts the one-dimensional possibilities. A trajectory along R can't retrace its steps else it would cross itself. Hence, if the point $x(t)$ gets moving at all, it must either keep going to infinity or else eventually slow down as it approaches some finite limit.

In the case of dynamical equations in two variables where the phase space is the plane R^2, there are further possibilities. A point's trajectory may be a closed cycle, with the point $x(t)$ periodically retracing its steps; and such a cycle may attract other trajectories in its vicinity. That is to say, other trajectories may loop round and round getting ever closer to a closed cycle (another trajectory can't actually join the cycle, because of the 'no merging' rule, but there can be asymptotic approach).

Attracting fixed points and cycles are our first, elementary, examples of *attractors* – i.e. bounded sets of points in phase space such that trajectories starting in their neighbourhood converge towards them. Attractors will feature crucially in what follows, for the long-term behaviour of trajectories (other than those that never go anywhere or those that fly off to infinity) are shaped by the attractors of a system. A nice topological result, the Poincaré-Bendixson theorem, shows that, since trajectories can't cross, fixed points and cycles are the *only* kinds of attractor that can occur in R^2 – which establishes that the long-term behaviour of a deterministic model which inhabits R or R^2 is very simple indeed. Trajectories can explode to infinity; or else be attracted to (or themselves *be*) a fixed point or closed cycle. All very tame.

And when we move to the case of (D)-style dynamical equations in three or more variables, the long-term behaviours of the trajectories they describe will still be relatively tame in the rather special case where we restrict the functions on the right of the equations to be simple linear functions of the state variables. However, as soon as we allow even a touch of non-linearity, things can suddenly get a *lot* more exciting. It is easy to see, in principle, that in the 3D case where the phase space is R^3 there might be room, so to speak, for trajectories to wind around in much more tangled ways yet still without crossing: but the new complexities we find in practice are utterly unexpected.

1.4 In a landmark paper (Lorenz 1963), Edward Lorenz investigated an extremely simple model of convection in the atmosphere. As air is warmed by the Earth's surface, a variety of convection patterns can be set up, of which the simplest consists of long horizontal cylindrical rolls (with the warmed air rising up one side and colder air falling down the other). Take a slice across the rolls, and we have in effect a series of two-dimensional rotational flows. Lorenz aimed to model one of these rotational flows. Starting from the classical equations for incompressible fluid flow, he ended up – after a whole raft of simplifying assumptions – with a cut-down system of ordinary differential equations in just three variables:

$$\text{(L)} \qquad \begin{aligned} dx/dt &= 10(y-x) \\ dy/dt &= 28x - y - xz \\ dz/dt &= xy - 8z/3. \end{aligned}$$

The exact significance of the three state variables here doesn't really matter for our purposes. But roughly speaking, x gives the speed of convection (positive values for clockwise flow, negative for anticlockwise), y is proportional to the temperature difference between the ascending and descending currents, and z indexes the shape of the vertical profile of the temperature distribution. The numerical values of the parameters – i.e. 10, 28 and 8/3 – are supposed to reflect plausible empirical values.

These equations are in the canonical form (D); the functions on the right are entirely well-behaved (e.g. continuously differentiable everywhere), so that a particular triplet of initial values will determine a unique set of values for every later time. The equations do not have a nice explicit solution; however, they can readily be numerically integrated by computer. And when he ran the numerical integration, Lorenz found that, for almost any initial state, the model soon settles down with the values of x, y, and z confined between definite limits. But within those limits the values vary in highly complex ways. For example, the value of x may first oscillate between various positive values (representing clockwise convection at changing speeds), with the oscillations getting larger and larger until the value of x overshoots the zero value and becomes negative. It then starts oscillating between various negative values (representing anticlockwise convection) until eventually being thrown back into a positive regime. The lengths of these alternating positive and negative regimes for x seem to vary quite randomly and unpredictably. And there are similar complexities with the other variables.

Lorenz also discovered (by accident!) that if he numerically integrated the equations from minutely different initial values then the values in the model after a relatively short time would be very different – the model, that is to say, exhibits a sensitive dependence on initial conditions.

It is easiest to grasp what is going on here if we again think geometrically, and plot 3D trajectories representing the evolving values of the state variables x, y and z for various initial conditions (Figure 1.3). At the outset, a trajectory may wander about, but if we follow it long enough then a trajectory – more or less wherever we start it – will end

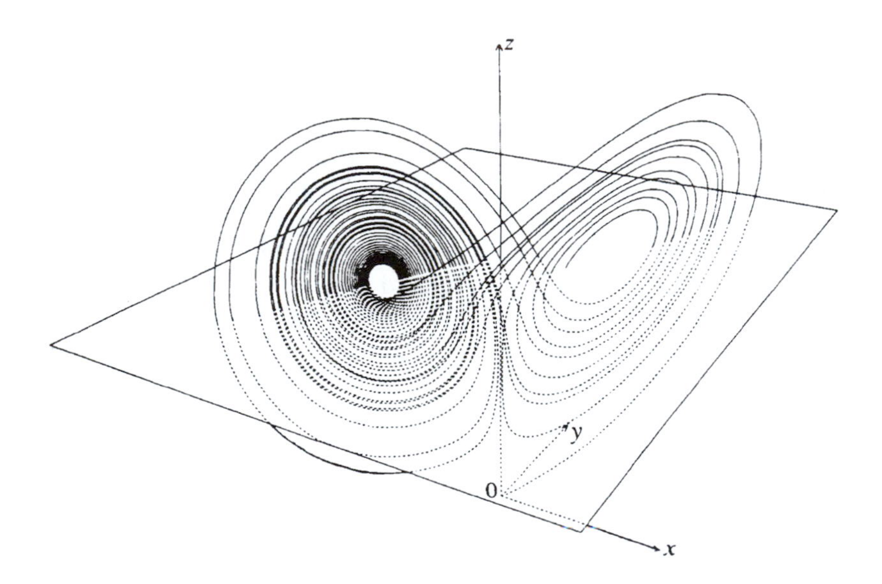

Figure 1.3 The Lorenz attractor

up winding around in a two-looped structure, asymptotically pulled in ever more closely towards the so-called *Lorenz attractor.* A trajectory spirals outwards around one wing of this structure until it gets far enough away from the centre to be thrown out of the spiral towards the centre of the other wing, from which it again begins to spiral outwards until it is thrown back to the first wing. The number of turns that a trajectory takes around one wing before jumping to the other is not fixed, however, but seems quite patternless. (Projected onto the x axis, this gives us the behaviour noted before – positive and negative oscillatory regimes alternating at what appear to be arbitrary intervals.)

Now suppose we take a pair of initial points that are close to each other and also near one of the wings of the Lorenz attractor. These points will initiate trajectories that will start spiralling around that wing; and the trajectories will diverge from each other as they go (very roughly speaking, doubling their gap on each circuit). Perhaps for a while they keep more or less in step as they get thrown together from one wing to the other. But eventually, the increasing divergence between the trajectories will mean that one of them is far enough out on a wing to get thrown across to the alternate wing whilst the other trajectory stays put for another circuit. This means that however close a pair of initial conditions are to each other, they will determine later values for the state variables that eventually won't even agree on a gross feature like the sign of x (i.e. won't even agree on the direction of convection). Micro-errors in fixing the initial conditions will thus inflate into macro-errors.

Since (L) is a deterministic system of equations, the trajectories they fix cannot intersect; so the flat appearance of the sketched attractor in Figure 1.3 is misleading – although the wings are nearly planar, there must be some depth in the structure to allow trajectories to cross from one wing to the other without actually intersecting. However, it also seems that a typical trajectory never precisely repeats itself (there aren't just a finite number of loops on each wing). The Lorenz attractor, then, wraps trajectories attracted to it into an almost flat-packed, never intersecting, bundle of infinitely long threads in which neighbours keep diverging. A *strange* attractor indeed.

Deriving the Lorenz equations

It is well worth pausing to indicate in general terms how Lorenz arrived at his equations. The exact details need not detain us, and we will be brisk and schematic; still, it is very important to realize the *kind* of simplifications and

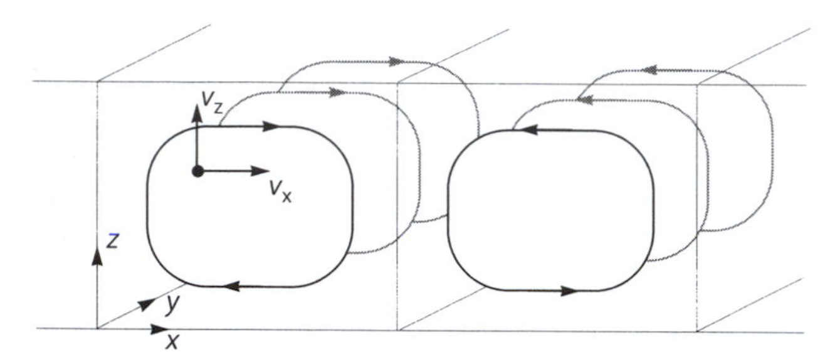

Figure 1.4 Rayleigh-Bénard convection rolls

approximations used, as they are typical in the sort of modelling work that yields chaotic systems of dynamical equations.

The target is an account of the flow in a convection roll in a layer of incompressible liquid between two planes with a constant temperature difference between the top and the bottom. It is assumed that there is no movement at all of fluid along the length of a roll, so that we can treat the flow as effectively two-dimensional ('Rayleigh-Bénard convection', see Figure 1.4: and that's simplification #1). We start therefore with a pair of standard Navier-Stokes equations for the remaining components (v_x, v_z) of the fluid circulation velocity at a point combined with equations for thermal diffusion. These equations contain a number of occurrences of a term for the fluid density, and this density will of course vary at different places in the fluid as the temperature varies. But we next cheerfully assume that the density can in fact be treated as *constant* except in one key term in the equations (this is the so-called Boussinesq approximation: simplification #2). The resulting equations can now usefully be rewritten in terms of a 'stream function' $\psi(x, z, t)$ and a temperature function $\tau(x, z, t)$, but still remain messy partial differential equations. So the next trick is to look for solutions which can be written as products of functions, each of which depends on only one of the variables x, z and t (simplification #3). And we'll assume that there's exact periodicity in the rectangular geometry, so that we are entitled to use 'Fourier analysis'. In other words, we assume that we can write the $\psi(x, z, t)$ and $\tau(x, z, t)$ as the products of infinite sums of periodic sine and cosine functions (simplification #4). But plugged back into the flow equations these infinite sums generate a correspondingly infinite set of ordinary differential equations, which isn't much use.

So to make further progress we need to throw away most of the terms in the Fourier expansions of the factors of $\psi(x, z, t)$ and $\tau(x, z, t)$. In fact, the proposal is to throw away *all but three* terms (simplification #5). This at last gives us some manageable differential equations: and ignoring one last term on the grounds that it isn't consonant with the previous simplification (which makes simplification #6) we end up with the equations (L) when empirical parameters are set plausibly.

Some of the simplifications used en route to (L) are moderately well motivated; but others – like the crucial and very dramatic truncation of the Fourier expansions – are simply pragmatic (justified if at all because what they deliver still seems to preserve some of the dynamical behaviour seen when somewhat less radically simplified equations are numerically integrated). Serious questions may be raised, therefore, about just how 'realistic' models like Lorenz's can be taken as being. What kind of explanatory weight, if any, can be given to such models? We will need to return to this issue.

1.5 The situation, then, is this. We have found that Lorenz's quite simple set of three interlocking differential equations dictates a strangely complex dynamics. There is considerable overall order: typical trajectories get pulled into the same small region of phase space around the attractor and then stay there, and so we can confidently predict e.g. their global shape or the speed at which neighbours diverge as they wind around the attractor. However there is radical disorder in the details. Typical trajectories don't repeat. There is no pattern in the length of the alternating visits which a trajectory makes to the two wings. Also there is sensitive dependence on initial conditions. Any error in fixing the initial state explodes exponentially so that we very quickly lose all fine-grained information concerning the later state of the system; we will know that it is somewhere or other in the vicinity of the attractor but not e.g. which wing it is on.

This type of combination of large-scale order with small-scale disorder, of macro-predictability with the micro-unpredictability due to sensitive dependence, is one paradigm of what has come to be called 'chaos'.

We won't attempt at this stage to give any more precise definition of chaos. That would be to proceed back to front: we would need to examine a lot more members of the family of simple-systems-with-complex-dynamics and compare their behaviours before we could be in a position to assess the claims of this or that technical definition to capture an interesting and 'natural' class of cases (see Chapter 10). So instead, let's look just a little more closely at the key ingredients of the Lorenz paradigm – the imposition of overall order by the existence of an attractor, and the simultaneous existence of sensitive dependence.

An attractor, as we informally introduced the idea, is a set of points in phase space such that all trajectories initiated in its neighbourhood converge towards it (as a limiting case, trajectories starting from points in the attractor itself will stay within the attractor). Define $S(t)$ as in Figure 1.2: i.e. $S(t)$ the set of points which trajectories from points in the set S reach at time t – so $S(t) = \{x(t) \mid x(0) \in S\}$. Then, we can a little more carefully define an attractor A as a closed set of points in phase space with the following properties:

(a) A is invariant under the dynamics – i.e. $A(t) = A$ for all t;
(b) there is some neighbourhood U containing A such that all trajectories starting in U are attracted towards A, i.e. if $x(0)$ is in U then the minimum distance between $x(t)$ and the nearest point in A tends to zero as t goes to infinity.
(c) A is *minimal*, i.e. no proper subset of A satisfies (a) and (b).

The largest such U containing all and only points initiating trajectories attracted by A is called the *basin of attraction* of A.

We have already seen that dynamical models don't have to have attractors at all – trajectories can e.g. all shoot off to infinity (it might be said that such trajectories *do* have an attractor, namely the 'point at infinity'; but this is not a point in phase space, and we have officially defined attractors to be subsets of phase space). On the other hand, even relatively simple dynamical models can have more than one attractor. Consider a pendulum again, but this time something more like a real clock pendulum, damped by friction and air-resistance, but – if it swings enough – also kept going by little kicks from the escapement mechanism on each swing. From many starting states, the pendulum will quickly settle down into a stable steady swing. Started from other states, which impart insufficient energy to get the escapement mechanism working, the pendulum will wobble about a little but then gradually come to rest. So a competent model of such a driven pendulum should allow for two attractors in the phase space, one a closed loop (representing the recurring path of the steady swing), the other a point (representing the rest state with zero displacement and velocity). And the phase space will divide into the basin of attraction of the steady swing and the basin of attraction of the rest state.

By definition, an attractor A pulls in bundles of trajectories towards it. Take a closed region of phase space S (within A's basin of attraction); over time, the dynamics takes S into $S(t)$, squeezed ever closer to the attractor. So we will expect the volume of $S(t)$ to be less than the volume of the original S. But while the volume may shrink, $S(t)$ may be more 'spread out' than S – as in Figure 1.2. This is what happens in the Lorenz case; a small ball of initial points S in the vicinity of the attractor will be shrunk in volume and 'rolled out' to a very thin, stretched set $S(t)$ wound round the attractor – as t increases, different points in $S(t)$ can end up far apart, near parts of the attractor on different wings. It is volume-shrinkage, roughly speaking, that gives us attractors: it is the spreading-apart of trajectories which start close to each other that gives us sensitive dependence on initial conditions. The crucial result illustrated by the Lorenz case is that we can have both together.

Various official definitions of sensitive dependence may be given. One standard definition looks like this: a dynamics is sensitively dependent (at a point x) if

$$\text{(SDIC)} \quad (\exists\epsilon > 0)(\forall\delta > 0)(\exists y)(\exists t > 0)(|x - y| < \delta \text{ and } |x(t) - y(t)| > \epsilon).$$

That is to say: sensitivity is a matter of there being some distance ϵ such that no matter how small a region around x ($= x(0)$) we take, there will always be some point y ($= y(0)$) in the region which gives rise to a trajectory which eventually gets more than ϵ away from the trajectory through x. However, this definition is very weak: it says nothing about *how fast* trajectories diverge; nor indeed does it say anything about *how many* points in a region around x will give rise to divergent trajectories (they could form a very sparse set of 'measure zero'). In typical chaotic cases like the Lorenz system something much stronger holds. *All* neighbouring trajectories diverge *exponentially fast* – at least to begin with (since trajectories are eventually confined to the neighbourhood of the attractor, the distance between a pair plainly cannot keep growing for all time). So we have, roughly,

$$\text{(EXP)} \quad |x(t) - y(t)| \approx |x(0) - y(0)|.e^{\lambda t}, \text{ where } \lambda > 0,$$

until the separation approximates the diameter of the relevant attractor. In appropriate cases, satisfying (EXP) obviously suffices for (SDIC).

Taken by itself, exponential error inflation is trivially easy to achieve. Consider the equation $dx/dt = x$, with the solution $x(t) = x(0)e^t$. An error δ in setting the initial value $x(0)$ inflates to an error δe^t as the system whizzes off to infinity. So error inflation by itself is entirely old-hat. The novelty in the new-fangled chaotic cases that will concern us is, to repeat, the *combination* of exponential error inflation with the tight confinement of trajectories by an attractor. As we noted, this pairing of micro-disorder and macro-order is impossible if the functions on the right in (D)-type equations are linear functions of the state variables. But once we turn to even quite simple non-linear examples like Lorenz's the combination turns out to be very common.

Incidentally, we can now see why the popular image of the Butterfly Effect really misses the target: for it only dramatizes the error inflation without at all reflecting the concomitant overall order that makes for a case of chaos.

Dissipative vs. Hamiltonian dynamics

It has been well-known since Poincaré's work on celestial mechanics a hundred years ago that even a very simple structure such as three bodies moving under mutual gravitational attraction can have a very complex dynamics. But this case, like most other cases studied by Poincaré and his immediate successors, is a closed dynamical system where total energy is conserved. Such conservative systems are standardly represented by Hamiltonian equations, which are nicely in the form (D), but with special constraints on the combination of functions that appears on the right.

Now, Hamiltonian systems are governed by Liouville's theorem – which says, in effect, that volumes in phase space are preserved as the system evolves. In other words, starting with a closed region phase space region S, the *volume* of $S(t)$ is constant. As already noted, preservation of volume is not preservation of shape: $S(t)$ can become more spread out, with a much larger maximum 'diameter', as time goes on – so Hamiltonian systems can still have a sensitive dependence on initial conditions. However, Liouville's theorem *does* rule out the existence of attractors: for as we noted, if an attractor A is to exist, then phase space volumes in its neighbourhood will be squeezed as the dynamical trajectories are pulled in towards A. (In a conservative case, a phase-space trajectory must stay confined to an 'equal-energy surface', i.e. to a region containing only states with a certain fixed energy level, and any other trajectory starting nearby with a slightly different initial energy will have to stay confined to its distinct energy surface.)

Conservative Hamiltonian systems of equations, then, cannot describe models with attractors; systems like Lorenz's where phase space volumes contract over time are cases where energy is continually dissipated. Systems of both kinds which exhibit a complex dynamics including sensitive dependence have been called 'chaotic'. But in fact, the conservative Hamiltonian cases and the dissipative cases are sufficiently different from each other for it to be quite attractive to recommend e.g. officially reserving the term *chaotic* for describing the complex behaviours of some dissipative systems, and using another term – *stochastic*, say – for the intricate behaviour to be found in the Hamiltonian systems. Be that as it may, our main concern in this book will be with dissipative systems with attractors as well as sensitive dependence. These cases are on the whole rather easier to understand and have in any case been the main focus of recent work.

But it is worth mentioning that the investigation of the stochastic wandering of Hamiltonian systems over equal-energy surfaces is crucial for e.g. the foundations of statistical mechanics. Consider the attempt to derive phenomenological gas laws from a conservative model of little billiard-ball molecules bouncing around in a box. This isn't as simple a matter as philosophical folklore sometimes suggests. Even if the molecules initially have motions represented by points randomly distributed in phase space, we might reasonably have expected that as the molecules keep colliding, correlations between the motions will be created, and the random distribution in phase space will be lost (we might imagine, say, that velocities will tend to get averaged out). However, in order to derive the gas laws, it is standard to assume that *at every instant* the phase space distribution is random – that there is, in Boltzmann's phrase, molecular chaos. And this assumption of continual re-randomization is not at all obviously consonant with the underlying dynamics for the collisions in the model. Doesn't the introduction of stochastic postulates over and above the deterministic Hamiltonian dynamics threaten an inconsistent overdetermination of the motions?

If the reduction of the gas laws by statistical mechanics is to go through, then, what needs to be shown is that the invoked Hamiltonian models *can* exhibit (most of the time) the right kind of stochastic behaviour, thoroughly mixing up trajectories so that phase space distributions remain effectively random. And this is where the concerns of chaos theory come into play. But this is a highly non-trivial matter, the topic of continuing research, and beyond the scope of this book.

1.6 We set out to say something about the kinds of mathematical model that chaos theory explores, and to say why recent work on chaos gives a new slant on the commonplace that small changes can have big effects. We now have the beginnings of a story.

Chaos theory is part of the general study of dynamical models, mainly concerned with the behaviour dictated by (especially non-linear) deterministic systems of equations in the canonical form (D). Which is why, as is often remarked, work on chaos can be of interest to theorists in many fields, not just physicists working in a variety of areas, but chemists, neurophysiologists, population biologists, and so on – indeed anyone who uses non-linear sets of equations that can be put in the form (D).

Of central concern are the frequent cases where we find the sort of combination of macro-order and micro-disorder which characterizes the Lorenz system. Consider the eddies in a stream, caused as the water rushes past a rock in the water. It is often the case that there is a quite stable pattern of – say – five eddies downstream from the rock (if you wade in and disturb the flow, the pattern disappears, only to re-establish itself when you return to the bank). However, the eddies continuously dance about slightly; and even when the pattern approximately recurs, the evolution of the pattern seems never quite the same. It might have been natural to suppose that such a combination of overall regularity with unpredictable fluctuations on the small scale must be either the result of horrible complexity (needing modelling as a very high dimension system) or else the result of continuous random noise perturbing some more regular behaviour, or both together. But we have now found that some dynamical systems can be low-dimensional and deterministic (so noise-free), and yet still generate this characteristic combination of behaviours. And *that* is the really novel discovery in recent chaos theory, and it opens up the possibility of tractable theories to accommodate what previously looked to be intractably complex behaviours.

So far, then, so good. But even the very introductory discussion up to this point already suggests a cluster of problems of a broadly philosophical kind.

Take the Lorenz equations again. The dynamical model – being sensitively dependent on initial conditions – tells us that very small factors can soon make sizeable differences; yet we have cheerfully thrown away many much more significant factors in the course of constructing the model by steps of radical simplification. So doesn't the model as it were undermine itself by showing that the simplifications that lead to it are illegitimate?

Moreover, the model's dynamics gets its character from the intricacy of the attractor, which makes trajectories for ever wind round, never repeating, never crossing, always spreading apart, in an infinitely tangled ball. These trajectories and their attractor may inhabit an abstract phase space; but the coordinates of points in this space are intended to represent physical quantities like the circulation velocity in a fluid. But it isn't at all clear that it makes physical sense to suppose that 'coarse-grained' quantities like this can be defined with infinite precision: so how can there be an infinitely intricate pattern in their time evolution?

These queries suggest a more general worry: how can such apparently unrealistic models as Lorenz's possibly *explain* anything? Maybe we can partially fend off such worries by saying that such a dynamical model may at least give us an *approximately* true account of certain phenomena: but what does the slippery notion of approximate truth really come to here?

Still considering the Lorenz model, we claimed that though the dynamics is deterministic it generates seemingly random behaviour – e.g. the sign of one of the state variables (the determinant of whether the convection is clockwise or anticlockwise) jumps from positive to negative at apparently random intervals. But is this mere appearance or is there really a sense of 'random' which *can* properly be applied to the output of deterministic processes?

These questions about intricate models and messy worlds, about truth and explanation, about determinism and randomness set the philosophical part of the agenda for the rest of this book. It might be protested that these issues, intriguing though they be, are not especially and uniquely raised by contemporary chaos theory. After all, questions about how precise mathematical models are to get applied to our messy world have been on the agenda since Newton's mathematization of physical enquiry three hundred years ago. Questions about how determinism can coexist with kinds of randomness are central to studies in the foundations of statistical physics (which go back at least a hundred years). Questions about approximate truth and about explanation have always been central to the philosophy of science. So, what's new?

Let's concede the point. Many of the conceptual and methodological problems that we are going to discuss are indeed problems outside the context of chaos theory. We would still face e.g. issues about the nature of approximative mathematical modelling even if specifically chaotic models had never been dreamt of. However, the philosophical problems are made particularly vivid by examples drawn from chaos theory. And it will be illuminating (in both directions, let's hope) to interweave their discussion with more explanations of the elements of chaos theory.

2

Fractal intricacy

2.1 Chaos – of the sort that concerns us – is a feature of certain dynamical models which exhibit sensitive dependence on initial conditions plus 'confinement' plus (typical) aperiodicity. Which is to say, roughly, that tiny differences in initial states can exponentially inflate into big differences in later states, but the values of the relevant state variables eventually remain confined within fixed boundaries although typically never exactly repeating.

Consider again, in the most general terms, how we can get sensitive dependence *and* confinement. Sensitive dependence means that trajectories through nearby points must tend to spread further apart from each other, confinement means that trajectories need to fold back on themselves so they keep within bounds. This kind of continual spreading and folding back is going to lead to a pretty tangled ball of possible trajectories. In fact, if there is typically to be no exact repetition, and also no merging or crossing of trajectories, then the ball must have a kind of infinite complexity. Since any given aperiodic trajectory will (eventually) be confined inside a finite region, it will have to keep revisiting some neighbourhoods of its earlier locations indefinitely often as it unendingly winds round and round. So segments of trajectories will have to be packed into such neighbourhoods infinitely densely. This is indeed what we seem to find when we use a computer to explore the ball of trajectories that wrap around the Lorenz attractor: if we blow up a small part of the attractor, we can discern more and more separate threads in that neighbourhood as we increase the magnification, apparently without limit.

This type of infinite intricacy – fine details within the fine details, ad infinitum – is characteristic of geometric structures which are *fractals*. And the interest of such complex fractal structures runs far beyond chaos theory: they are intriguing in themselves and there are other putative applications. So this chapter introduces some elementary ideas from fractal geometry. We also discuss what sort of role fractals can possibly have in the description of a physical world which we know not to be infinitely intricate in the relevant respects. We only return to chaos proper at the end of the chapter.

2.2 The term 'fractal' is a relatively recent coinage, introduced by Benoit Mandelbrot; but some of the intricate mathematical monsters the term describes have been well-known to pure mathematicians for a century or more. We will begin by considering a hackneyed but approachable example.

Start with an initial straight line K_0. Cut out the middle third of K_0 and replace it by two lines the same length as the removed segment to make the kinked line K_1 (as in Figure 2.1). Now perform the same operation on the four straight parts of K_1, replacing each middle third. That yields the crinklier line K_2 made up from sixteen shorter straight segments. Perform the same operation on the middle thirds of these segments to get K_3; and then keep on going an infinite number of times. The limit of this sequence of operations, K_∞, is a mathematically well-defined set of points, the *Koch curve*.

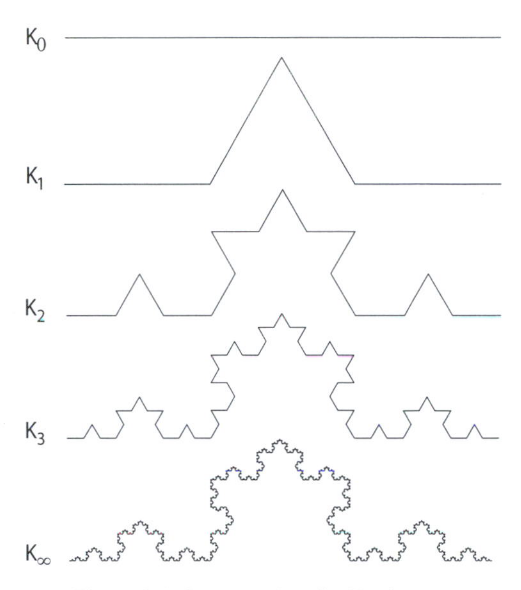

Figure 2.1 Constructing the Koch curve

Three initial remarks about this construction. (1) We can equally well think of the transformation from K_0 to K_∞ as involving not cutting-and-replacings but bending-and-stretchings of the original straight line (without tearing or crossing or self-touching). And any such continuous distortion of a line can be thought of as a one-dimensional 'curve'.

To expand on that, note that in characterizing a geometric figure an important set of descriptions are the *topological invariants* – i.e. the descriptions that continue to apply to the figure (set of points) even if it is stretched and distorted without tearing. Being a certain *size*, say, isn't a topological invariant; contrast being *simply connected* (being all of one piece, rather than consisting of two discrete portions), which *is* an invariant. Now, the notion of dimension – as we'll soon see – is many stranded: but *topological dimension* is technically defined to meet the intuitive, commonsensical, conditions that (a) the definition always yields an integer value, (b) as expected, sets of disconnected points in Euclidean space come out as zero-dimensional, ordinary lines and curves as one-dimensional, ordinary surfaces are two-dimensional, solid figures three-dimensional, and so on, while (c) the dimension number of a figure is topologically invariant (stretching doesn't affect dimension). The technical details don't matter: but conditions (b) and (c) together imply that K_∞ must come out as topologically one-dimensional.

(2) K_∞ is self-similar. Take e.g. K_2 and shrink it to one-third size; put four copies end to end joined at the correct angles; and you get K_3. Likewise, take K_n, shrink to one-third, join four copies together, and that produces the next crinkliest curve, K_{n+1}. And at the limit, if you take K_∞, shrink it and then join four copies, you get ... K_∞ again (since K_∞ is already as crinkly as possible). In other words, the Koch curve can be seen as consisting of four similar parts, each an exact one-third scale replica of itself.

(3) K_∞ is infinitely long. Assume that K_0 has unit length. K_1 has length 4/3. Each further construction stage increases the length of the line by the same factor 4/3. So K_n has length $(4/3)^n$; and that length goes to infinity with n. Of course, the as-the-crow-flies distance between any two points on the curve stays finite; it can be no more than the length of the initial line K_0. What goes to infinity is the distance traversed walking-along-the-curve-following-all-the-ins-and-outs. A nice corollary is this: if you take three Koch curves and join them end to end in a triangular shape to make a 'Koch snowflake', then the result will bound a finite area – but this finite area will have an infinitely long (because infinitely crinkly) border.

To pursue the question of length a bit further, suppose we want to estimate the length of an ordinary curve using a pair of dividers. Set the gap between the points of the dividers to be δ, and then count n_δ, the number of δ-long steps that fit along (with the points of the divider staying on the curve). The length of the curve is approximately given by $n_\delta \delta$ – and the smaller δ is, the better the approximation. For a normal curve, if we (say) halve the divider gap δ, the number of divider steps along will approximately double. More generally, as we go towards the limit of smaller and smaller divider gaps δ, reducing δ by a factor of f will increase n_δ by a factor of f.

Measuring a curve's length by dividers is in effect taking equidistant points located on the curve, joining them with straight lines, and then measuring instead the length of this kinked line. Suppose you want to measure the area of some surface in an analogous way. The procedure would be to take a triangular grid of points on the surface, each δ from its neighbours. This defines a net of (flat) equilateral triangles, and the total area of this net is the area of a δ-sided triangle multiplied by the number of triangles n_δ – i.e. $(\sqrt{3}/4)n_\delta\delta^2$. Again, the smaller δ is, the better this will approximate to the area of the original surface. Halve the gap δ between points and the number of triangles in the corresponding net will approximately quadruple. And more generally, for a normal smooth surface, in the limit of smaller and smaller δ, reducing δ by a further factor of f will increase n_δ by a factor of f^2.

We will not worry here about higher-dimensional analogues; let's just take the power law already suggested by our two initial cases. In the limit as δ tends to zero, the effect of changing the gap δ between the equidistant checkpoints along a normal curve or surface is given by

(D) Proportional increase in n_δ = {reduction factor in $\delta\}^d$

where n_δ is the number of point-joining elements and d is the dimension of the figure to be measured.

Consider, however, what happens when we use a pair of dividers on the Koch curve. Start with a unit opening, and the dividers can be laid off against K_∞ just once. Now reduce the divider gap δ to one third and we can lay off the dividers against K_∞ exactly four times. When we set the gap to 3^{-n}, then 4^n steps are needed (3^{-n} is the length of each segment of K_n, so it is as if we are stepping from vertex to vertex along that kinked line). Further decreasing δ by another factor of three makes n_δ increase by a factor of four; this scaling behaviour persists as δ tends to zero, and is in fact independent of the initial divider opening. So, plugging these values into (D), we get

$$4 = 3^k$$

where k is the 'dimension' of K_∞. Solving,

$$k = \log 4/\log 3 \approx 1.262.$$

One response to this odd result might be to say that it just goes to show that (D) no longer applies to mathematical monsters like the Koch curve. But it is better to stipulate that there is a notion of 'dimension' – *divider dimension*, let's call it – which relates to the way that measurements at different scales behave and which by definition obeys (D). Then K_∞ is a one-dimensional figure in the ordinary, topological sense, but it has a divider dimension strictly greater than one: and this splitting apart of the different dimension measures is a useful indicator of 'monstrosity'.

Now, the notion of divider dimension is suggestive, but it is of somewhat limited application. As we will soon see, there can be monstrous but 'gappy' sets, such that we can't step along them with fixed divider settings while staying inside the set. And there are other limitations too. So let's immediately introduce a second notion of 'dimension', which is again related to the way measurements behave at different scales, but which has much more general application.

Imagine a figure drawn on a piece of graph paper with squares of size δ. We can indicate the size of the figure by counting up the number b_δ of the squares which contain portions of the figure. As the mesh-size δ decreases, the number of 'boxes' occupied by parts of the figure will increase. If the figure is a normal curve, halving the box size will roughly speaking double the number of occupied boxes; and more generally, b_δ will tend (in the limit) to go up by a factor of f when the δ decreases by f. Likewise if the figure is a plane area bounded by a normal curve, then b_δ will tend (in the limit) to go up by a factor of f^2 when δ decreases by f. The obvious power law has the same pattern as before: in the limit as δ tends to zero,

(B) Proportional increase in the number of occupied boxes b_δ
= {reduction factor in box size δ}^{d}

where d is the dimension of the figure.

Now, suppose we try 'box-counting' the Koch curve. Imagine K_∞ drawn on graph paper with mesh δ, with b_δ boxes occupied. If K_∞ were shrunk to one-third scale, and the mesh size were simultaneously reduced to $\delta/3$, there would still be b_δ boxes occupied. But now recall that K_∞ already comprises four one-third scale copies of itself (Figure

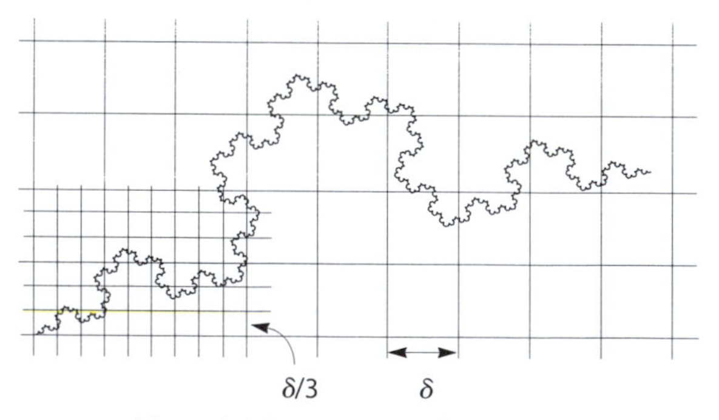

Figure 2.2 Box-counting the Koch curve

2.2): hence K_∞ must occupy approximately $4b_\delta$ boxes of side $\delta/3$ ('approximately', to allow for the fact that the four portions of K_∞ are at different angles and so will intersect with the grid somewhat differently). As δ gets smaller, further reduction of δ by a factor of three will lead to a proportional increase in b_δ which gets closer and closer to four. Applying (B) gives us $4 = 3^k$, where k is the 'dimension' of K_∞; so again $k \approx 1.262$.

As before, we could say that this merely indicates that (B) no longer applies to mathematical monsters. But let's instead stipulate that there is another notion of dimension – *box-counting dimension* – which also relates to scaling behaviour and which by definition obeys (B). Then we have just shown that K_∞ has box-counting dimension strictly greater than its topological dimension.

Note that the notion of box-counting dimension extends smoothly beyond the initial 'graph paper' conception to cover the case of a figure S embedded in an n-dimensional space. Take the space to be partitioned by a grid of n-dimensional boxes of side δ – when $n = 1$, the 'boxes' are just intervals – and count the number b_δ of boxes which intersect with S; work out how, in the limit, reducing the size of the boxes alters the number of boxes which intersect S. Then solve (B) for d to give the box-counting dimension of S.

2.3 Let's quickly consider two more introductory examples. First, suppose we again start with a unit line C_0 and punch out the middle third, this time without replacement, leaving C_1, which comprises two closed intervals [0, 1/3], [2/3, 1]. Then we punch out the middle thirds of these intervals in C_1, and so on ad infinitum (Figure 2.3). The points that remain at the end form C_∞, the 'middle-thirds' Cantor set (another much-visited old exhibit in the zoo of mathematical monsters).

C_∞ is a totally disconnected 'dust' of points, containing no

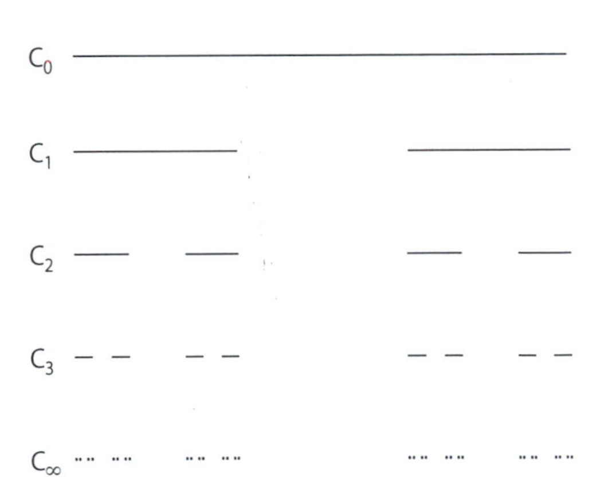

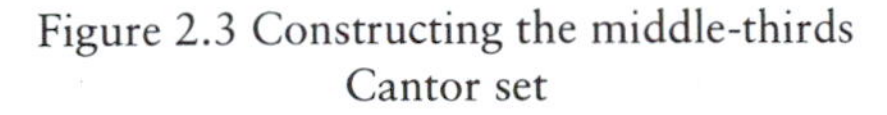

Figure 2.3 Constructing the middle-thirds Cantor set

intervals. At the first construction stage, an open interval of length 1/3 is removed; at the next stage, two intervals of length $(1/3)^2$ are cut out, then four intervals of length $(1/3)^3$, and so on. The total length removed is thus

$$1/3.(1 + (2/3) + (2/3)^2 + \ldots) = 1$$

Hence the remaining length of C_∞ is zero. However, this dust – maybe rather surprisingly – has the cardinality of the continuum. (Why? Imagine the real numbers between 0 and 1 being given by *base three* expansions – note, though, that some numbers do not have unique representations, since e.g. .0222... = .1 . Then the move from C_0 to C_1 in effect knocks out the numbers which must have '1' as the first digit in a ternary expansion. The move from C_1 to C_2 knocks out the remaining numbers which must have '1' as their second digit; and so on. So C_∞ comprises all the numbers which have ternary expansions in just '0' and '2'. But there are as many of *these* expansions as there are expansions in just '0' and '1' – i.e. continuum many.)

What about the dimensions of C_∞? Dusts like this have the same 'ordinary' (topological) dimension as any other sets of isolated points, namely zero. Divider dimension isn't defined (the set is too 'spaced out' to step along from point to point with fixed divider settings without falling into the gaps). But the box-counting dimension *is* defined, and is easily calculated. We set C_0 to have unit length; so C_∞ will fit into a unit interval 'box'. Reduce the box-side to 1/3, and you'll need two boxes (one for the left, one for the right half of C_∞); reduce the box-side to 1/9, and you'll need four boxes to cover C_∞. And so forth. Applying the rule

(B) Proportional increase in the number of occupied boxes b_δ
= {reduction factor in box size $\delta\}^d$

we get

$$2 = 3^c$$

where c is the box-counting dimension of the Cantor set, and thus

$$c = \log 2/\log 3 \approx 0.631.$$

Like the Koch curve, the box-counting dimension of the Cantor set is therefore strictly greater than its topological dimension.

The Cantor set is also like the Koch curve in being self-similar. For each n, C_{n+1} comprises two one-third scale copies of C_n. So at the limit, C_∞ comprises two one-third scale copies of itself. However, this kind of strict self-similarity is the exception rather than the rule among the intricate monsters of the Koch and Cantor families. Consider, for instance, a variant Koch-like construction where at each stage there is randomization; in effect, a coin is tossed to decide whether the kink replacing a middle third is to be set above or below the line. Infinitely iterating this randomized construction procedure yields a curve RK_∞ (Figure 2.4). There will be zero probability that e.g. the first quarter of RK_∞ is an exact scale replica of the whole; so we lack strict self-similarity. Yet any RK_∞ is as infinitely prickly as the self-similar K_∞, and its scaling behaviour is the same. So the divider and box-counting dimensions of RK_∞ will be the same as for K_∞.

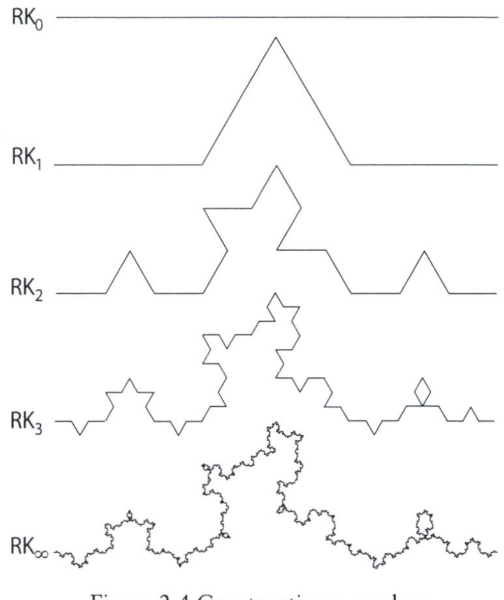

Figure 2.4 Constructing a random Koch curve

The described ways of generating Koch curves and the middle-thirds Cantor set illustrate two common ways of constructing fractals, namely by repeated replacements and by repeated subtractions. And there are myriad different ways of improvising on these same themes. For example, in the Cantor-like cases, we could e.g. divide intervals into five at each stage and remove the second and fourth portions; or the divisions could be done asymmetrically and/or randomized. In the Koch-like cases we could add flatter or steeper kinks; or the kinks could involve more than two added lines; or could be added asymmetrically, etc. Moreover, it is only for simplicity's sake that our examples so far have involved constructions starting from a straight line. We could start e.g. with a square; divide it into a three-by-three grid of nine smaller squares, and take out the central one. Now do the same for each of the remaining eight smaller squares – divide up, and remove the middle ninth. And keep going … Or start with a pyramid, paste smaller pyramids onto each triangular face, then repeat the recipe ad infinitum. The variations are endless, and we can in these ways in fact customize monsters of arbitrary box-counting dimension to order.

2.4 We have informally introduced two numbers that can be associated with certain figures or sets of points – divider dimension, and box-counting dimension. The use of the term 'dimension' for these quantities is of course not unmotivated: but it *is* a stretched usage of the term compared with the 'ordinary' topological notion (and that's why there is nothing incoherent about supposing that such dimension measures can turn out to be fractional). The divider dimension of a set of points is quite often not defined; divider dimension and box-counting dimension can take different values for the same set even when they are both defined; and there are other complications. So henceforth we'll mostly concentrate on applications of box-counting dimension.

It can readily be shown that the box-counting dimension of a figure cannot be *less* than the topological dimension. But it can be *greater*, as we saw in the cases of the infinitely intricate Koch curves – plain or random – and the Cantor set. This divergence of dimension measures is typical of intuitively monstrous constructions, and motivates a working definition. A *fractal* monster, we will say as a first shot, is a set whose box-counting dimension exceeds its 'ordinary' dimension.

Note however that it is only e.g. the final limit curves K_∞ and RK_∞ which are true fractals in this sense. This is crucial in what follows. 'Prefractals' such as RK_3 or RK_{303} (say) are perfectly ordinary finite-length lines made up of 4^3 or 4^{303} straight segments – their divider and box-counting dimensions are exactly *one*.

Take divider dimension: once the divider gap δ is set to be much shorter than the straight segments of RK_{303} (i.e. once $\delta \ll 3^{-303}$), then the dividers will just 'see' a series of straight lines occasionally changing direction. Further reduction of δ by, say, a third will therefore triple, more or less, the number of times the dividers can be laid off against RK_{303}. Thus, in the limit as δ tends to zero, reducing δ by a further factor of f will lead to f more divider lengths fitting along. Hence by (D) RK_{303} has divider dimension one.

True, if you measure RK_n using a comparatively large divider gap δ (i.e. with δ much longer than one of the straight segments of RK_n), then reducing δ by a factor of three will enable the dividers to be laid off against RK_n about four times as often; and similarly for some further reductions. Hence there is fractal-like anomalous measurement behaviour of RK_n at coarse scales. But *only* at coarse scales. Once the divider gap δ is much shorter than one of the straight segments of RK_n, then we merely have common-or-garden, one-dimensional scaling.

Exactly likewise for box-counting dimension. Place a coarse grid over RK_n and count the number of occupied boxes: refine the grid by a factor of three and there will be roughly four times as many boxes occupied. So that's fractal-like scaling behaviour at coarse scales. But the rule (B) defines box-counting dimension in terms of what happens *in the limit*, as we reduce box-size to zero. And once box-size is reduced below the size of the straight elements of RK_n, further reduction of box-size by a factor of f will just make for roughly f times as many occupied boxes – one-dimensional behaviour again.

The Cantor case is the same. The limit set C_∞ is a fractal by our definition; but any C_n, for finite n, is simply the union of 2^n short intervals, which makes for a perfectly ordinary object which is one-dimensional in all applicable senses.

To conclude. The zoo of fractal monsters contains many even queerer beasts, generated in other ways we haven't touched on yet (including that modern icon, the Mandelbrot set). But for the present, we don't need to explore much further. We have seen that fractals have a certain infinite intricacy, with fine detail at *every* scale. But if the detail gives out, so that at a sufficiently small scale we have a collection e.g. of ordinary intervals, as with the prefractals RK_n or C_n, then the box-counting dimension will equal the 'ordinary' dimension, and by definition we don't have a fractal after all. And that already raises an obvious question: can *infinitely* intricate fractal structures, as opposed to mere prefractals, possibly have a role in the description of a non-intricate nature?

Fractal dimension

A glance at a standard text (e.g. Falconer 1990, §2.2) shows that the canonical notion of dimension used in fractal geometry is actually that of *Hausdorff-Besicovich* dimension, which is different again from box-counting dimension. And a fractal is usually officially defined as a set of points whose Hausdorff-Besicovich dimension exceeds its topological dimension. Why introduce this further notion?

Consider the set Z of rationals in the interval [0, 1]. This, being a disconnected set of points, has topological dimension zero. But its box-counting dimension is *one*. Since the fractions are spread evenly along the unit line, we need a non-gappy covering of interval 'boxes' to include all the rationals in Z. Reduce the size of the covering boxes by a factor f, and we'll simply need f times as many. Which counts, by (B), as one-dimensional behaviour. Worse still, consider the very sparse subset of rationals

$$Z^* = \{0, 1, 1/2, 1/3, 1/4, \ldots, 1/n, \ldots\}$$

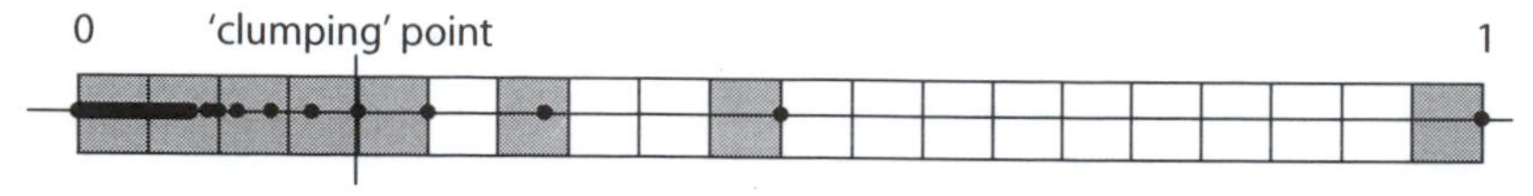

Figure 2.5 Box-counting the set Z^*

which again has topological dimension zero. But its box-counting dimension is as large as 1/2. (Proof sketch, purely for fun! Set the box-size to δ. The larger members of Z^* will get their own boxes (Figure 2.5), while the smaller ones will be clumped together in shared boxes in a continuous run. The clumping starts once the distance between $1/k$ and $1/(k+1)$ starts falling below δ, i.e. when $1/k.(k+1) \approx \delta$. Above this clumping point, the fractions $\{1, 1/2, 1/3, \ldots 1/k\}$ each have their own box, which makes k boxes. Below the clumping point there are as many boxes as are needed to cover the interval $[0, 1/k]$, i.e. approximately $(1/k)$ divided by the box-length δ, making another k boxes. So in all, we need about $2k$ ($\approx 2/\sqrt{\delta}$) boxes of side δ to cover the set Z^*. It is immediate that as we further reduce δ by some factor f, we need a factor of $\sqrt{f}$ more boxes to cover Z^*. Hence by the definition (B), the box-counting dimension of Z^* is 1/2. QED)

That's decidedly embarrassing. Remember, the definition of a fractal aims to capture an interesting class of mathematical monsters; and we offered, as a first shot, divergence between topological dimension and box-counting dimension as a mark of monstrosity. But now we find that even Z^*, which is as tame and domesticated as an infinite set can possibly be, has a box-counting dimension strictly greater than its topological dimension. Our criterion of fractal monstrosity has thus proved far too lax.

So it's back to the drawing board! What we need is a definition of fractal dimension that somehow ignores the little bits of grit thrown into the works by 'spread out but very sparse' sets like Z^* which overlap with lots of boxes and so give misleadingly high box-counts. The classic Hausdorff-Besicovich definition does the trick: but we need not worry about *how* it does the trick – rather fortunately, as the definition's cost in complexity and difficulty of application is rather high. In most 'nice', ungritty, cases (especially the types of cases that come up in physical applications) the Hausdorff-Besicovich dimension of a set is equal to the more tractable box-counting dimension. So for our practical purposes, we can continue to emphasize the latter.

Finally, why didn't we say that a fractal is a set whose box-counting or Hausdorff-Besicovich dimension is *fractional* (as the very label 'fractal' might suggest)? Because monstrous fractals can have integral dimension. (Proof sketch, again just for fun. Consider the Broken Koch curve BK_∞ constructed as in Figure 2.6. Now adapt the argument by which we found the box-counting dimension of the original K_∞. Take a covering of b_δ boxes of side δ; a one-third scale BK_∞ will then occupy b_δ boxes of side $\delta/3$. But BK_∞ is self-similar, comprising three one-third scale copies of itself; so (roughly) BK_∞ will need $3b_\delta$ boxes of side $\delta/3$ to cover it. The reduction factor in δ thus equals the increase factor in b_δ. Which gives BK_∞ box-counting dimension *one* (it also has the same Hausdorff-Besicovich dimension). Yet the Broken Koch curve is a disconnected dust of points like the Cantor set, containing no intervals, so it is zero dimensional in the topological sense. It thus has fractal dimension greater than its topological dimension, and so counts as a fractal, even though all its dimension measures are non-fractional. Which seems intuitively right, given the infinite intricacy of BK_∞. QED)

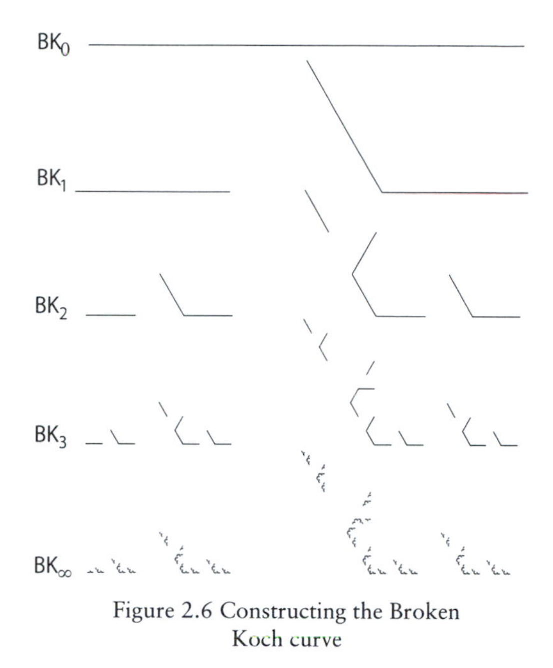

Figure 2.6 Constructing the Broken Koch curve

2.4 The basic mathematics of fractal sets is not new; what *is* relatively new is the suggestion that these bizarre fancies, originally dreamt up by pure mathematicians for their own entertainment, might be useful for describing a range of natural phenomena.

Consider first Mandelbrot's famous discussion in 'How long is the coast of Britain?' (Mandelbrot 1967; see also Mandelbrot 1983, §5). Suppose you set out to measure the length of the western half of the coast of Britain. Take a detailed map, and a pair of dividers. Set the dividers to measure (say) steps of 200 km, and see how many such steps you can lay off between John O'Groats and Land's End, with divider points always landing on the coast. Now set the dividers to measure 100 km; this will force you to follow the coast a little more closely, and the measured length of the path will increase. Set the dividers more finely, and you have to now track round more bays and promontories. As you follow the crinkly west coast more and more accurately, the measured length will continue to increase. Let δ be the distance represented by the gap between the divider points; let n_δ be the number of steps along the west coast. Then it supposedly turns out empirically, that over the range of divider lengths representing 200 km to 1 km, we have roughly the following rule:

(D*) Proportional increase in n_δ = {reduction factor in δ}$^{1.2}$

We might thus be moved to say that the coastline is a fractal pattern of (divider) dimension 1.2. And indeed, it looks really rather like the random Koch curve RK_∞ (Figure 2.4) whose dimension is only slightly greater.

Three points about this. (a) The coastline may be something *like* a random fractal, but it can't really be one – for the idea that the coastline has detail at every scale is absurd. Even if we conventionally agree that the line is defined by, say, mean high-water mark, there is no sense in trying to fix that even to the nearest millimetre, let alone at all finer scales. This isn't just an epistemic point about what can be known; the claim is that the idea of an absolutely precise high-water line is *physical nonsense*, given what we know about shifting sands, the way water molecules get sprayed about, etc.

In fact, (b) the coastline is no more like a random fractal akin to RK_∞ than it is like some construction akin to RK_n for suitable n. As we stressed before, if you use long enough divider steps in measuring along the prefractal RK_n then you will also get anomalous scaling behaviour at coarse scales (in the sense that closing up the dividers will initially increase the number of needed steps by a non-integral power-law). But this doesn't make RK_n a fractal. Why should it be any different for the coastline? Mandelbrot writes that

> ... within the scales of interest to the geographer, coastlines can be modelled by fractal curves. Coastlines are fractal patterns.

However, it would be at least as correct to say that within the scales of interest to the geographer, coastlines can be modelled by certain *prefractal* curves. Unlike the genuine fractals, the prefractals lack all the extra infinitely intricate detail with no empirical content; and if anything, modellings that lack redundant content are to be preferred when available. Of course, since coastlines aren't fully determinate, the claim can't be that a given coast precisely instantiates a certain prefractal (rather than a fractal): whether we model using prefractals or fractals, there's some idealization either way. The claim is simply that prefractal modellings must be *at least* as good as fractal ones; so we can have no reason to say that coastlines are fractal patterns.

(c) Even if (over a certain range) measurements of the length of the west coast do scale quite regularly with divider length in a fractal-like way, why suppose that this is more than a mildly diverting accident? The gross structure of the coast depends on the way the underlying rock strata are folded and tilted: but finer structure will depend on (say) the way that the exposed surface rocks – often of a quite different type – are eroded by rivers and the sea. It will presumably involve a good measure of luck if the gross structure and the finer structure exhibit sufficient similarity to render true an interesting coarse scaling power-law. Again, a geological map reveals that the rocks that shape the coast of the Northwest Highlands are quite different from those of Cornwall: so again it will be luck if any scaling law applies (even approximately) right down the coast.

The truth about the coastline of western Britain, then, is really rather unexciting. By happenstance, it is roughly like one of those finite, non-fractal, constructions of the type RK_n – constructions which are produced by an iterated operation which, *if* carried out infinitely often, *would* produce a fractal. No reason here, therefore, to talk of the 'fractal geometry of nature' – or not, at any rate, if that is meant literally.

Take another example. The arterial trees of healthy kidneys are more complex that those of diseased kidneys. The relative complexities can be numerically indexed by box-counting X-ray pictures and working out the coarse scaling ratio, and the results can be diagnostically significant. Still, this scaling behaviour would only give a true fractional box-counting dimension if it persisted in the limit – which of course it doesn't. Renal arterial trees aren't fractals: their fine detail is limited.

The moral generalizes. Given a surface-descriptive appeal to fractals, where it is claimed that some natural phenomenon has a fractal-like look to it, we can agree that there may well be anomalous measurement behaviour at coarse scales. There may, in this sense, be a (pre)fractal-like geometry to aspects of nature. And at least some of the time this seems to be all that Mandelbrot, in his more careful episodes, is claiming (see e.g. Mandelbrot 1983, 18, where he says 'no one believes that the world is strictly ... scaling'). We should certainly be cautious about leaping to assert that some department of nature really has a fractal geometry in any stronger sense. For we always need to ask: won't a *better* description of nature in fact be provided by prefractals that lack the infinite excess detail? To which the answer seems invariably 'yes'.

Barnsley ferns and fractal growth theory

To reinforce the last point, let's consider another type of example where it may initially seem compulsory to deploy fractal geometry in describing certain worldly phenomena.

By way of introduction, consider the original Koch curve K_∞ again. We characterized the curve as the result of an infinite process of replacing middle thirds; but there are other ways of characterizing the same curve. Imagine that we have a sort of idealized xerographic copying machine that can produce a number of reduced copies of the original image on the same sheet. For example, the machine may have four lenses so arranged that we get four one-third scale copies of an original image aligned in the now familiar basic Koch array. Call this a Koch copier (Figure 2.7). If we now take a suitable initial image, and Koch-copy it, and then Koch-copy the result, and then Koch-copy the new result, and so on, the result at each stage will get nearer and nearer to being like the Koch curve. Koch-copy a 'nice' initial image an infinite number of times, and the result will be exactly K_∞. And Koch-copying K_∞ itself will give us precisely K_∞ again. Indeed we can *define* the Koch curve as that unique set of points which is invariant when Koch-copied.

Now, we can readily imagine a whole family of variant machines, with different numbers of adjustable lenses, and where the lenses allow not only reduction and rotation of images, but also allow reduction by different degrees in different directions, mirror reflection, and the skewing of images. We might call a multi-copier of this more general type a *Barnsley copier*, in honour of Michael Barnsley who has investigated such copiers (or, more precisely, the 'iterated function systems' which the copiers instantiate: see

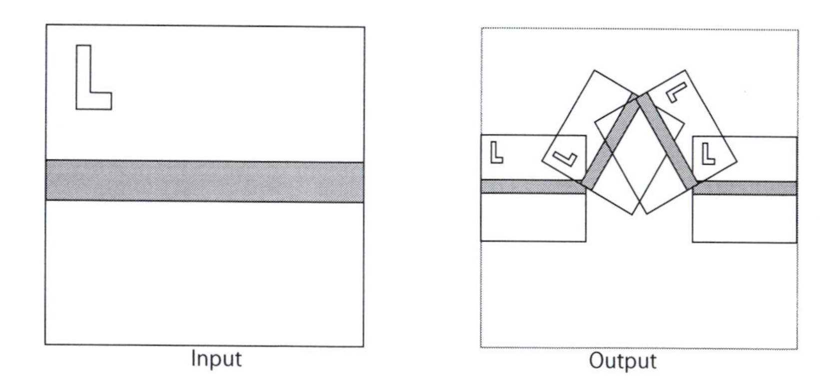

Figure 2.7 The Koch copier
(The 'L' serves as an orientation marker)

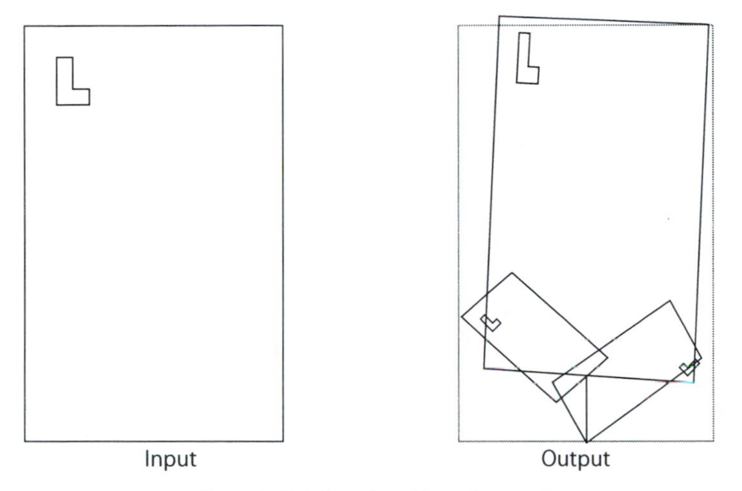

Figure 2.8 A four-lens Barnsley copier

Barnsley 1988). When an image is fed into a Barnsley copier, and then the result fed through again, and so ad infinitum, the result is typically a fractal (a fractal which is invariant under copying, and whose box-counting dimension can be readily calculated from information about the disposition of the lenses).

Consider a particular case – the four-lens copier whose effect is illustrated in Figure 2.8 (one lens produces an *extremely* thin rectangle, showing up as the near-vertical line towards the bottom of the output; the other three lenses produce skewing and in one case reflection). If the initial input is, say, a completely black 'page', then on repeated iteration the output is one of the rather startling icons of fractal geometry, reproduced overleaf (Figure 2.9) so as not to spoil the surprise!

It is difficult not to be *very* struck by this result, and feel that the fractal character of the resulting Barnsley fern must yield some deep insight into the structure of ferns themselves – somehow, this fractal seems more quintessentially fern-like than many a fern! But again, we should resist being too impressed. Agreed, some real ferns exhibit a limited degree of rather approximate self-similarity over about four length scales (a whole frond is like one of the branching sub-fronds which is like one of its component leaflets which is in turn rather notched). But four levels of approximate self-similarity is hardly infinite, fractal self-similarity. And note that our four lens multi-copier has 24 different numerical parameters to tweak (in order to specify the operation of each lens we must set six parameters: two coordinates for the centre of the resulting image, two scaling factors, a rotation angle and a skewing angle);

Figure 2.9 The Barnsley Fern

we also need to set four Boolean parameters, to determine which lenses produce mirror reflections. With that many parameters to play with, maybe the fact that we can run up a custom Barnsley fern that looks quite like one of those four-level self-similar ferns should, on second thoughts, be no great surprise. Moreover, the genuinely fractal Barnsley fern B_∞ – the result of *infinitely* iterating the copying procedure, with all the redundant fine detail – will be rather *less* like a real fern than is some prefractal B_n produced after a dozen or two copies (it is, after all, a finite computer-generated prefractal that is reproduced in Figure 2.9).

The moral seems to be as before. A natural phenomenon may have the look of some interesting *pre*fractal. But that must fall far short (indeed, *infinitely* far short!) of establishing that the phenomenon in question reveals nature to have a genuinely fractal geometry.

Unlike the case of Mandelbrot's coastline, there may be an interesting story to be told about what systematically generates the limited and approximate degree of self-similarity that we find in some ferns. This will presumably be a complex matter of cell-biology, not yet very well understood. In other much simpler cases, however, we can readily produce accounts of the growth of (approximately) self-similar structures. This is the province of so-called fractal growth theory, which deals with matters as varied as the spread of epidemics in a population and the accretion of electrodeposits. Since this is perhaps the main body of theory (apart from chaos theory) where ideas from fractal geometry are invoked, let's briefly pause to ask what real work the ideas are doing here.

Take a very simple example: if you place a carbon cathode in the middle of a suitable dish of liquid surrounded by a zinc ring plate, and run a current through, then you'll get a rather fern-like growth of zinc metal 'leaves' spreading out from the cathode. And this growth will have a certain fractal-like

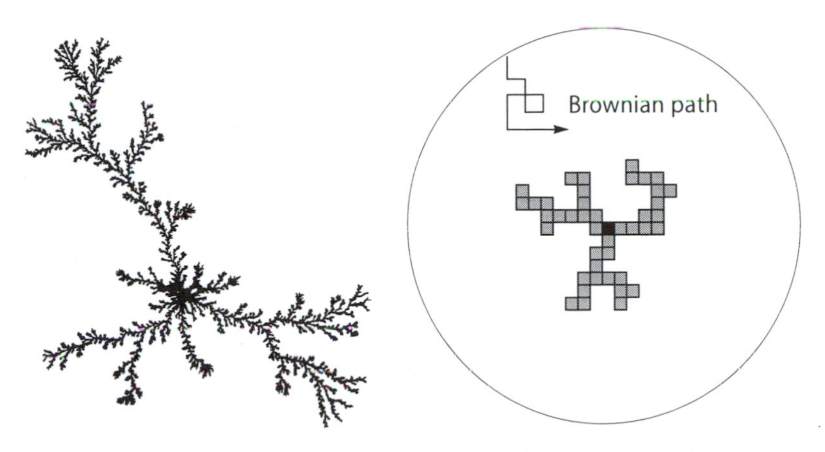

Figure 2.10 Fern-like accretion – and a simple Brownian model

character, with the main branches growing smaller fronds which in turn grow yet smaller accretions, etc. (Figure 2.10)

What is going on is that zinc ions are released from the surrounding plate, and wander along random Brownian paths until they happen to hit the fern-like growth and stick onto it. We can construct a simple model of this growth as follows. Take a square mesh, draw a circular boundary and fix a single 'sticky' particle in the centre. Now inject a new particle somewhere at the boundary and allow it to jump randomly from square to adjacent square. If and when it occupies one of the four squares immediately next to the original sticky particle, it stays put and becomes sticky itself. Keep injecting new particles (from anywhere on the circular boundary). These will wander about until they reach a square next to a sticky particle, and then attach and turn sticky. The consequent growth of the sticky chains of particles will have a fern-like structure. By tweaking features of the model (e.g. by starting with an hexagonal mesh rather than a square one, or by making stickiness probabilistic, and so on) we can simulate different kinds of accretion patterns.

There are two points to note about this sort of modelling. First, this type of probabilistic accretion model leads to fractal-*like* structures. We can demonstrate e.g. that box counts at different coarse levels of detail will have fractal-like scaling properties (and we can compare models to reality by comparing the predicted degrees of 'crinkliness' with the crinkliness of the real growth phenomena as measured by empirical box-counting at different scales). So it is rather *as if* we are dealing with fractals with non-integral box-counting dimensions. But second, our model is in this case not a truly fractal model at all – it can't be, as there isn't limitless detail *even in the model*: the 'ferns' in the model are by hypothesis built of little units without further structure.

The real content of various fractal growth theories like this is again conceptually quite unproblematic (and I should hasten to add that scientists working in the area are usually entirely clear-headed about the point being emphasized). It can't really be said that the relevant natural structures are fractals, or even that they are best modelled by truly fractal structures with infinite intricacy; the claim is only that they can be modelled by finite structures with coarse scaling features which, *if* pushed to infinity, *would* make them fractals.

2.5 General issues about fractal geometry and its possible applications, as we've seen, have some stand-alone interest. But our concern in this book is with relevance to chaos theory. We noted that to get sensitive dependence, confinement and aperiodicity, trajectories will need to wrap around each other in an infinitely intricate way. If the trajectories are being pulled in ever closer to an attractor, then that means that the relevant attractor will need to be correspondingly intricate. We will expect, therefore, that the 'strange' attractors for chaotic dynamical systems will have a fractal geometry.

And that's certainly what we seem to find in the Lorenz system, for example – 'seem' because we still lack watertight formal proofs. Of course, the Lorenz attractor is not produced by iterated simple additions or deletions like the Koch curve or Cantor set; but computer trials apparently reveal unending depths of structural fine detail as we magnify portions of the attractor. Box-counting at various scales using a computer and also more abstract mathematical analysis suggest that the Lorenz attractor is indeed another fractal object – this time with box-counting dimension ≈ 2.07.

For more about this fractal object, see Chapter 4. Given what's gone before, however, the obvious first question to raise is: can a fractal object like this *really* play an essential part in the description of nature? The discussion so far might make us generally wary of such a suggestion. The next chapter, however, aims to defuse such premature scepticism about the very idea of applied chaos theory.

3

Intricacy and simplicity

3.1 In the opening chapter, we saw that the chaotically complex behaviour in the paradigm Lorenz case is dictated by a 'strange' attractor with an infinitely intricate structure. In thc following chapter, we were able to sharpen up that informal talk of infinite intricacy; the attractor, we said, is a *fractal*. We must now face the question which immediately arises. How can an infinitely intricate structure like this possibly play an essential part in a competent scientific account of some natural phenomenon? For by the lights of our own best physical theories, quantities such as fluid circulation velocity, temperature, the proportional concentration of a chemical in a mixture and so forth – that is, macroscopic quantities of the type dealt with in paradigm chaotic models like Lorenz's – cannot have indefinitely precise real number values. Hence their time evolutions cannot really exemplify infinitely intricate trajectories wrapping round a fractal attractor, any more than a coastline can exemplify a genuinely fractal pattern.

What is being claimed here is something much stronger than the trite epistemological point that there is a limit to the precision with which we can know facts about the values of physical quantities. The claim is that *there is no fact of the matter* about the exact values of quantities like circulation velocity. We know, for example, that fluids are gappy distributions of molecules in motion. So, the circulation velocity of a fluid at a point P can be nothing other than the average velocity of all the molecules of the fluid contained in some small ball centred on P – or better, it might seem, the true circulation velocity is to be identified as the *limit* to which this average velocity tends as the size of the small ball around P shrinks to zero. But the trouble with the latter suggestion is, of course, that once we shrink the ball around P to the scale where it only contains a few dozen molecules, further shrinking will produce increasingly wild fluctuations due to the random Brownian motions of the molecules – and (in all probability) the final limit of the average velocity of the molecules in a very small ball around P will be zero because there will be no molecules at all left in the ball. To avoid this absurdity in 'taking a limit', the best we could do is average the velocity of molecules in some arbitrarily selected standard volume (for example, a sphere of radius one micron centred at *P*). But even this still wouldn't produce perfect determinacy. For example, quantum indeterminacies on the usual understanding entail that there is no exact fact of the matter about e.g. which molecules have their centre of mass determinately in the ball at a particular time. In short, there seems no principled way of completely precisifying the quantity *circulation-velocity-at-a-point*.

Parallel physical arguments suggest that the temperature at a point *P* in a fluid – determined by, inter alia, the average kinetic energy of the molecules in some ball centred at *P* – must have a parallel lack of perfect precision. But if the velocity and temperature at a point are not perfectly determinate, then derived quantities such as mean circulation velocity or temperature gradient can't be perfectly determinate either. Yet these are the very kinds of quantity that e.g. the Lorenz model aims to represent.

For another case, consider the concentration of a chemical in a mixture. The interest of this case is that a much-studied example of an apparently chaotic phenomenon in nature is the so-called Belousov-Zhabotinskii reaction. This involves a linked set of reactions between four chemicals which can manifest itself in sudden and very dramatic colour changes in the mixture, occurring at apparently random intervals. You can keep the reaction going for an indefinite period by setting up the apparatus so as to feed a continuous stream of new reactants into the tank of stirred chemicals: and we seem to observe some overall order in the colour switches combined with never-repeating changes in the detailed dynamics – the combination characteristic of chaos. It is not hard to develop some plausible, well-motivated, mathematical models of the reactions which have the appropriate chaotic features. The values of the variables in such a model will, as always, be real numbers (that is to say, indefinitely precise real numbers) – yet these numbers represent chemical concentrations, and it makes no physical sense to suppose that the chemical concentrations themselves are indefinitely precise quantities. Apart from quantum indeterminacies about the precise moment when a molecule of one chemical ceases to be, and a molecule of some other reactant comes into being, there are quite boring indeterminacies concerning e.g. the precise moment at which some new molecule from the input stream of reactants counts as getting into the mixture.

This sort of argument can be repeated with modest variations for most if not all the gross physical quantities represented in paradigm chaotic dynamical systems like Lorenz's. Note, the claim is not that *every* physical quantity is necessarily imprecise. Take, for a simple example, the ratio of the masses of the electron and the proton – this unit-free quantity may very well have a perfectly exact value, determinate to an unlimited number of decimal places. At any rate, there is no obvious objection of principle to this suggestion. (After all, we readily assume that the ratio of the masses of two electrons, or the ratio of the charges of an electron and a proton, are both *exactly* unity – determinately so, with quite unlimited precision – so why shouldn't other ratios be equally determinate?) The current argument, to repeat, only concerns various 'coarse-grained' macro-quantities. But where it applies, there cannot be infinitely intricate fractal patterns in the time evolution of the quantities in question; and so chaotic models which postulate such an infinite intricacy must, it seems, necessarily misrepresent the facts.

To summarize: we initially noted that

(F) The chaotic behaviour in models like Lorenz's depends on trajectories getting pulled ever closer to a strange attractor with a fractal geometry.

It has now been argued that

(G) The evolving physical processes that chaotic dynamical models like Lorenz's are characteristically intended to represent cannot themselves exhibit true infinite intricacy.

And (F) and (G) together imply the conclusion that, at least in the typical case, the very thing that makes a dynamical model a chaotic one (the unlimited intricacy in the behaviour of possible trajectories) can not genuinely correspond to something in the time evolutions of the modelled physical processes – since *they* can not exhibit sufficiently intricate patterns at the coarse-grained macroscopic level.

3.2 The conclusion that typical applied chaotic theories must, in a sense, misrepresent the facts looks secure. For (F) is not in dispute. Maybe there is some room to argue about particular cases of (G). For example, it was suggested in passing that the temperature at a point P in a fluid is inexact because the average kinetic energy of the molecules in some ball about P will not be precisely determinate (whether we take limits, or fix on a standard sized ball). And some might protest that the argument here rests on a resistible reductionist assumption, namely that temperature in fluids is constituted by mean molecular kinetic energy. However, we need not pause to argue the particular case. For there are plenty of other instances of (G), like the chemical concentration example, which don't rest on any contentious reductionist assumptions – more than enough to combine with (F) to yield the implication that there will very typically be a certain mismatch between infinitely intricate chaotic dynamical models and a messier world.

The issue, then, is not whether applied chaotic models will usually have more structure than there can possibly be in the world being modelled, but rather how far (if at all) this should be thought of as a problem. Are we, for example, forced to conclude that all talk of chaos in the world as opposed to our models indicates the same excess of rhetorical zeal that we found in talk of coastlines as being fractal patterns?

Here's a first-shot response:

> It should already be a familiar fact that theoretical models regularly idealize and so misrepresent the facts. For example, when starting classical mechanics, we learn to treat a pendulum as a simple harmonic oscillator. But this pretends that the oscillations are vanishingly small, that the suspension point is ideally rigid and frictionless, that air resistance is zero, and so on and so forth. Yet these admitted idealizations don't mean that the classical theory of the pendulum is useless. In some good sense it helps explain, for example, why the period of a pendulum is more or less independent of the angle of swing, and explains the relationship between pendulum length and period. Again, on a grander scale, we can usefully deploy e.g. 'barotroptic' models of the atmosphere where wind velocity doesn't change with height. Such a model radically misrepresents the facts, but the results are illuminating none the less. So to be sure, there is a general issue about what is going on when we deploy idealizing mathematical models. But then there is no *new* problem raised by the fact that applied chaotic dynamics must idealize too.

This response, however, seems rather too quick. Let's agree that applied mathematical models regularly depart from strict truth. But in the familiar old-style case, such as the pendulum example, the departure consists in abstraction from some of the physical details – we ride roughshod over complications, hoping that they don't matter too much (and perhaps trusting that we can cope with them later if necessary by adding refinements to the basic modelling equations). By contrast, the immediate worry about the chaotic dynamical models with their infinite fine structure is that they have too much detail rather than too little; they are opulent rather than impoverished. The difference between familiar models and chaotic models, it might be suggested, seems to be rather like the difference between a small scale map that ignores a lot of the on-the-ground detail (abstracting in order to produce a comprehensible overview) and a map with an unlimited amount of excess, necessarily fictional, content. How can using the *second* kind of map be legitimated?

It will be protested that the contrast just drawn is factitious.

> Most mathematical models which aim to represent the evolution of macro-quantities will necessarily have surplus content, since the models use perfectly precise mathematical apparatus to represent what have just been argued to be somewhat indeterminate properties of the world. A Newtonian model, for example, may predict perfectly elliptical motion for a perfectly defined point centre of gravity. But even this familiar case is replete with idealized surplus content. For a start, there are no infinitely precise point centres of gravity in nature (the position of the centre of gravity of a planet, for example, is as indeterminate as the answer to the question whether this or that speck of local dust counts as part of the atmosphere). Surplus content almost always comes with the use of precise theoretical models: so after all there is no novelty in the chaotic cases.

Again, however, this reply seems too easy. Presented with a modelling problem, our task usually involves, inter alia, trading off accuracy in hitting the empirical data against structural simplicity and economy. There is no problem in seeing how the best solution balancing these conflicting demands may be a model deploying e.g. point masses and perfect ellipses (that way, we keep the basic model clean and simple, while perhaps allowing some slack in the way it gets matched to the coarse-grained, fuzzy-valued, quantities in the world). But the question now is: how can the best solution to a modelling problem involve a structure with infinite intricacy of the kind characteristic of chaos? How can models involving unlimited and necessarily non-empirical fine structure ever count as being *simple* enough to be the best trade between hitting the data and structural economy? Merely noting that perfect ellipses, say, are also idealizations doesn't even begin to hint at a useful account of what makes for simplicity in the chaotic cases or why it might be useful to idealize nature by models involving wildly intricate fractal structures.

It will immediately be said, though, that we don't have to look very far for the needed account of what makes for simplicity in cases like Lorenz's:

> We saw in Chapter 1 that Lorenz's dynamical model, for all its intricately complex structure, is specified by three very simple linked differential equations in three variables. Indeed, this is the very thing that is remarkable about the Lorenz example – that such a very simple set of equations should characterize a mathematical structure with such complex behaviour. It turns out, in other words, that relatively simple sets of equations can in fact specify infinitely intricate solutions as easily as they can specify e.g. nice elliptical solutions. We need look no further, therefore, for an account of where the simplicity is to be found in the familiar paradigms of chaotic systems: it's a matter of being governed by simple equations.

There no doubt has to be something right about this response; but *what* is right about it needs teasing out. After all, it certainly is not the case – as the Lorenz case itself so vividly shows – that superficial simplicity of equations (in the sense of having a low number of variables, small number of terms on the right of the differential equations, etc.) goes along with virtues like ease of solution: thus there is no immediate pragmatic gain in simplicity of equations. So then why should we care about the surface simplicity?

To press the point. A given set of dynamical equations specifies a structured bundle of allowed trajectories through some phase space. The possible time-evolutions of a real-world dynamical phenomenon can also be mapped by the possible trajectories of some point representing (with some arbitrariness) the slightly fuzzy state of the system. The prime criterion for a mathematical model to count as providing a good match to the real-world phenomenon is then, roughly speaking, that the two bundles of trajectories, the predicted and the actually possible, in some way track each other closely. (For some elaboration of this rough story, see Chapter 5.) From this perspective, what gets compared with the world is not, in the first place, a system of equations but rather the structures for trajectories which the equations specify. However, if it is the *structures* that matter, why value simplicity of *equations*? Unless, that is, simplicity of equations is a mark of intrinsic simplicity of described structure. But we need now to say a bit more about why that is so, and about where the intrinsic simplicity is to be found in the fractal trajectory structures found in the chaotic case.

Let's pause to consider what emerges from the to-and-fro of the dialectic in this section. We noted that applied chaotic models will (typically) misrepresent by having infinitely intricate surplus content. We can live with this, treating it as just another case of the way idealizing theories depart from strict truth, *if* we can find some compensating virtue – roughly, some story about simplicity to trade off against the empirical mismatch. Superficial simplicity of equations *per se* doesn't seem to fit the bill: we at least need some story about how simplicity in equations relates to *intrinsic* simplicity in the described dynamics. Spelling out that story is the next task.

3.3 Return to the Koch curve again. Start with K_∞ and stretch it by a factor of three in each direction. Now fold it back on itself, with the axis of the fold a bisecting line at right angles to the baseline (see Figure 3.1). The result is a smaller Koch curve with a different orientation. Now fold *this* back on itself, with the axis of fold a bisector at right angles to the new baseline. This second fold takes us back to the original K_∞. The Koch curve is invariant, then, under a certain combined *stretch-and-fold* operation. Moreover, suppose we take those axes of fold to be fixed in advance (rather than being specified relative to a given curve); then it is fairly easy to see that a particular K_∞ will be the only set of points which is an invariant of the combined operation of stretching-and-folding-round-those-fixed-axes.

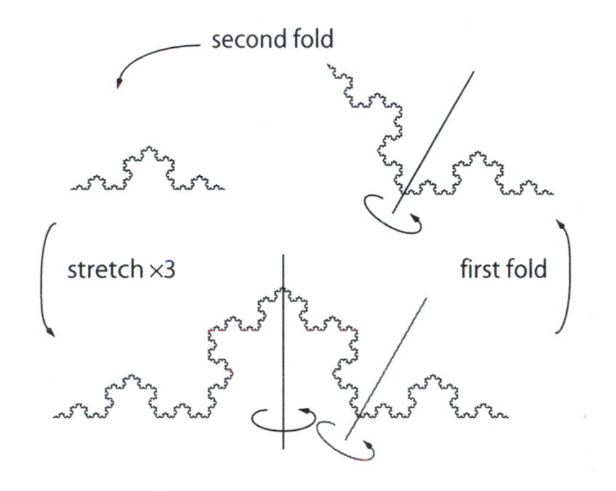

Figure 3.1 Stretching and folding the Koch curve

This observation points to another way of characterizing some fractals, as sets of points that remain invariant under certain stretch-and-fold operations. (This is closely connected with thinking of certain fractals as invariants under Barnsley copying – see the §2.4 interlude 'Barnsley ferns and fractal growth theory'. Reverse the sequence of operations in Figure 3.1 and you get shrinking and then the production of copies joined at various angles, just as in a Barnsley copier.) But now recall again the point made at the beginning of the last chapter: we can get chaos – in particular, get confinement of trajectories plus sensitive dependence on initial conditions – if phase space trajectories get trapped into a finite region by an attractor, and so fold back on themselves, while locally they keep spreading apart. In other words, *chaos results when the dynamics of the system functions to stretch and fold the bundle of possible trajectories within a certain region.*

That last idea is very schematic, but it is the key to understanding much about chaos. At various points (starting with the next interlude on 'Stretching and folding' and §4.2) we will explain and illustrate it in more detail. But for the moment, let's ask: inside the trapping region, what will be invariant under this sort of stretch-and-fold dynamics? Answer: the attractor for the system. For remember, the attractor is defined to be invariant under the dynamics (while any off-attractor bounded set S of points in phase space is not invariant since it gets pulled in by the dynamics to the set $S(t)$ closer to the attractor and hence distinct from S – see §1.5).

So now the following comparison suggests itself. If we think of the Koch curve as produced by an infinite sequence of cuttings-and-replacings, then we naturally think of its intricacy as a kind of monstrous complexity. But switch perspectives, and think of the curve as what is left invariant by a very simple set of operations, and a kind of intrinsic simplicity comes back into focus. Similarly, if we stare at the infinite detail of e.g. the Lorenz attractor, we naturally think of it as an astonishingly complex object and then wonder how such a mathematical monster can legitimately get put to empirical work ('where's the simplicity to compensate for all the surplus content?', we asked). But switch perspectives again, and think of the attractor as what is left fixed in place by a dynamics which stretches and folds phase space trajectories, and we now can see how the needed simplicity might get into the picture. For we could have a dynamical model which specifies relatively simple stretching-and-folding operations, yet (as the example of the Koch curve shows) even very elementary stretches and folds can have infinitely intricate fractal invariants.

In other words, there could be modest, common-or-garden, idealization at the level of our story about the way the dynamics kneads the trajectories, and yet a fractal attractor might (as it were) still drop out as a consequence of that simplifying story.

And how does this relate to simplicity of equations? Well, consider again equations in the canonical form

(D) $dx_i/dt = F_i(x_1, x_2, \ldots x_n)$.

The F_i directly specify how the x_i change over time, and those rates of change fix the way that bundles of trajectories spread apart, bunch together, fold back or whatever. Relatively simple F_i can mean simple spreadings or bunchings or foldings etc.: *that* is why simplicity in equations matters.

Stretching and folding

It will help to fix ideas if we pause straight away to consider just a little more carefully how a dynamics can 'stretch and fold' trajectories.

As already noted, dynamical models like Lorenz's have two key features – phase space volumes contract rapidly over time and trajectories starting from afar get pulled in towards a finite trapping region. Proof of contraction (trapping is equally quickly demonstrated): If we have a system of n equations in the canonical form (D) – i.e. equivalently

$$dx/dt = F(x)$$

where x is an n-vector – then it is a standard result (Strogatz 1994, 312) that

$$dV/dt = \int_V \nabla \cdot F dv$$

where V on the left is the volume of some region of phase space, and the integral on the right is taken over the interior of that region. In the case of the Lorenz equations (§1.4), a simple calculation shows that $\nabla \cdot F = -c$ (where c = 41/3). Hence

$$dV/dt = \int_V -c dv = -cV$$

and so

$$V(t) = V(0)e^{-ct}.$$

Thus, in the Lorenz model, volumes in phase space not only shrink, but shrink exponentially fast with time. QED.

So, even if we start off with a large ball of phase space points S, the dynamics will eventually map points in S to points in some *very* much smaller volume set $S(t)$ contained in the trapping region. To put it another way, the bundle of trajectories starting from points in S gets squeezed together over time.

But it is more complex than that. For in the Lorenz case we find that while phase volumes keep contracting, this overall shrinkage is produced by very fast compression in one direction transverse to local trajectories, and a much slower spreading apart in the other orthogonal direction. See the schematic Figure 3.2a. The effect therefore is that bundles of trajectories, once inside the trapping region, rapidly get squeezed into something like a thin sheet, in which the trajectories more slowly spread apart, as if the sheet is being

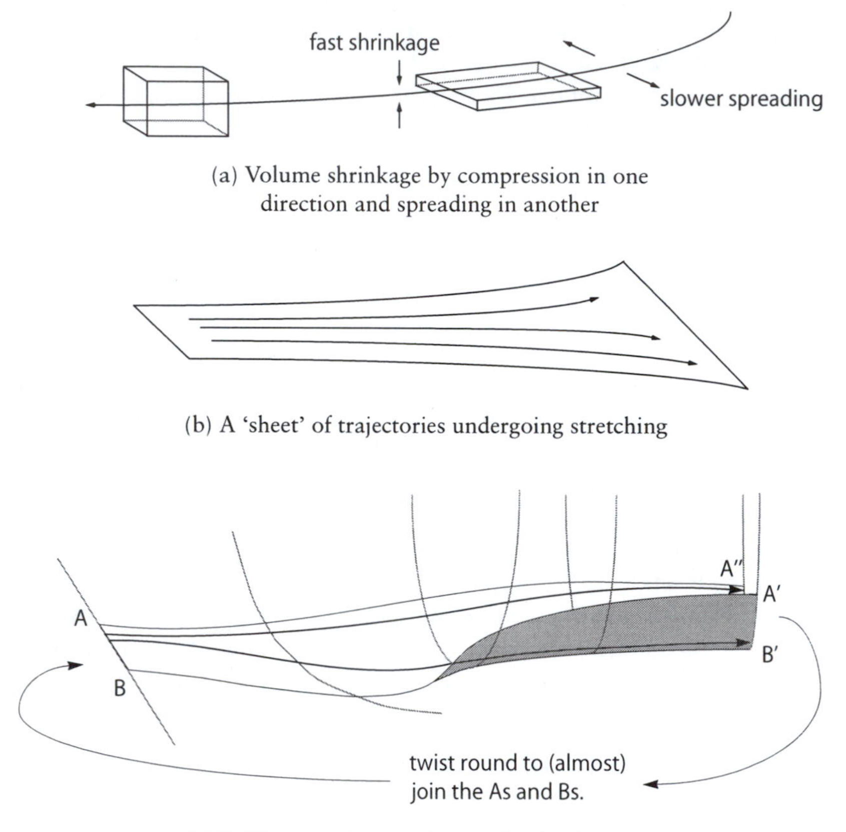

(a) Volume shrinkage by compression in one direction and spreading in another

(b) A 'sheet' of trajectories undergoing stretching

(c) Folding a trajectory sheet – the simplest case

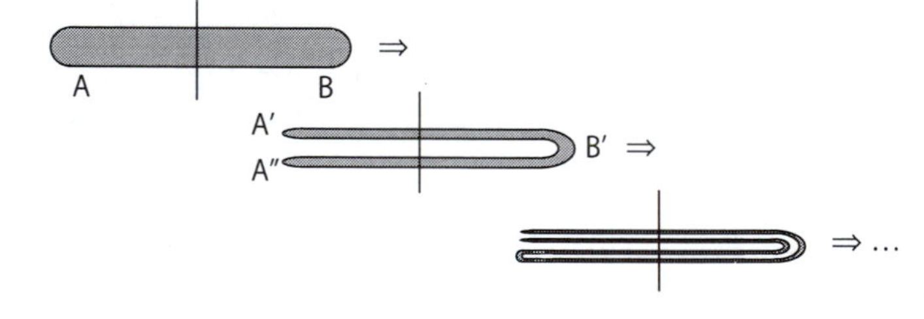

(d) Horseshoe folding

Figure 3.2 Stretching and folding trajectories

stretched. See the equally schematic Figure 3.2b. This dual effect is rather common in other dynamical systems too.

Now, to stay confined in the trapping region, this 'sheet' needs to fold back on itself somehow. A very simple kind of folding is illustrated in Figure 3.2c, with the result sketched in Figure 3.3. Note that the folded 'sheet' here cannot be an ordinary surface exactly rejoining itself, or else the trajectories drawn on the surface would repeatedly have to intersect each other, contravening the 'no-crossing' rule for deterministic systems. The 'sheet' must have some thickness then. So, as a bunch of trajectories starting within a thin layer along AB twists around to almost rejoin, the effect must be rather as illustrated in Figure 3.2d – after one circuit the original thin layer gets folded into a compressed horseshoe shape (of no greater depth), and then after another circuit this horseshoe will get folded into a sort of double horseshoe, and so on. But now consider a cross-section through the layer (i.e. along the verticals in Figure 3.2d). After one circuit a middle portion of the layer gets removed; after another circuit, the middles of the remainders get cut out, and so on – the now familar construction of a Cantor set. So the ultimate effect of this kind of folding is to pull the trajectories into what we might dub 'Cantor pastry' – i.e. the trajectory bundles are (in the limit) flattened out into sheets packed closely like leaves of filo pastry, but with the cross-section being a Cantor-type set. A fractal structure results.

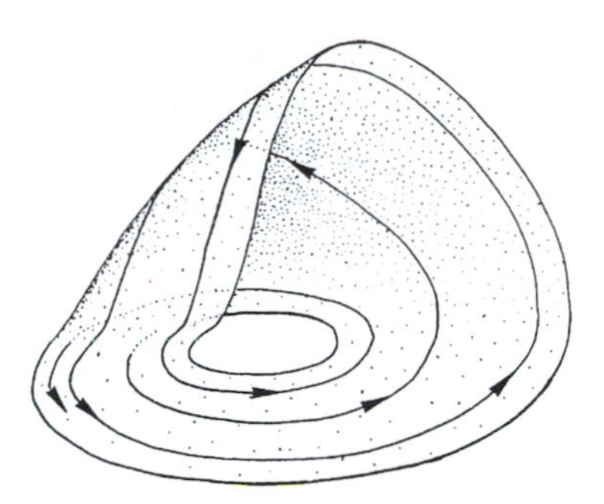

Figure 3.3 A schematic Rössler band

To sum up. We started with the combination of simple features that we find in e.g. the Lorenz case – (a) trajectories ultimately confined into a trapping region, and (b) overall volume shrinkage going with spreading in one direction transverse to trajectories, thus compressing trajectories into thin layers. We then postulated a bit more structure by considering the simplest imaginable way of securing confinement of these spreading layers (i.e. by folding trajectory sheets round onto themselves). And out of this story about the simplest stretching and folding of trajectory-sheets there has already dropped a fractal attractor.

Of course, the imagined single-fold structure sketched in Figure 3.3 cannot be quite what is involved in the original Lorenz case: however, as we will see in the next chapter, only a very modest variation on the basic idea is needed to get back to the familiar two-winged geometry of the Lorenz attractor. For the moment, we'll finish by simply remarking that a single-fold attracting 'band' structure (Figures 3.2c, 3.3) *is* realized more or less exactly in some other much investigated cases, e.g. the so called Rössler system governed by the equations

$$\begin{aligned} dx/dt &= -y - z \\ dy/dt &= x + ay \\ dz/dt &= b + z(x - c) \end{aligned}$$

for suitable values of the parameters.

3.4 We wondered: 'How *can* the best solution to a modelling problem involve a structure with infinite intricacy of the kind characteristic of chaos?' Now we see the makings of an answer: the fractal intricacy of an attractor may (in a sense) *come for free* when we have simplicity and modest idealization elsewhere in the model.

In other words, suppose we concentrate on modelling the way that the dynamics of some phenomenon basically stretches apart and folds back trajectories (trajectories, remember, of state-representing points in some suitable phase space). We will no doubt precisify and simplify, but at this stage in an entirely standard way. However, despite the simplicity here, a fractal may yet drop out as the invariant attractor. And unlike the cases discussed in §2.4, where talk of fractals involved some rhetorical excess, it may indeed be a true fractal that is integral to the model. When describing the coastline or a fern, we argued that a non-fractal will do just as well: by contrast, if we are to preserve the simple stretching and folding of the dynamics, then we will have to live with the concomitant fractals as the attractors. (We will expect such structurally simple dynamics to be specifiable by superficially simple equations, like Lorenz's. However, to repeat, the thought is that it is the structural rather than the superficial simplicity that fundamentally matters.)

Hence our best applied dynamical theories, conforming to normal canons of simplicity (including minimization of surplus content, etc.) may yet, in seeing a certain elementary overall pattern in the way trajectories develop, *ipso facto* discern Lorenz-like chaos. Agreed, in thus discerning worldly chaos, dynamical theories must strictly misrepresent the facts – but, it now seems, in no more problematic a way than other idealizing theories.

4

Predictions

4.1 The argument of the last chapter raises as many questions as it answers. Chaotic models will idealize and misrepresent the facts by involving patterns of dynamical behaviour which have an intricacy that the modelled phenomena must typically lack. Agreed, other theoretical models also misrepresent by idealizing and simplifying: and we argued that chaotic models may not be dramatically worse off in this respect. Still, it might be wondered, just how comforting is the thought that chaos theory is in the same unseaworthy boat as other theories?

Perhaps we can cheerfully live with the fact that chaotic theories won't be strictly true if we can make sense of the natural fall-back position, namely that these theories may yet be *approximately* true. But is the notion of approximate truth in good enough shape to rely on here (the literature is littered with failed philosophical analyses)?

That's business for Chapter 5. First, however, there's a more immediate issue. We suggested that a dynamical model can represent the stretching and folding of trajectories in a way which meets standard canons of simplicity and yet as a result be chaotic. But chaos involves sensitive dependence on initial conditions, meaning that the inevitable errors in setting initial conditions inflate exponentially, which in turn means (it seems) that the practical predictions of a chaotic model must go *spectacularly* wrong. The question then seems to arise: what kind of 'modelling' can there be when there must be spectacular and pervasive predictive error? How can a model possibly be both chaotic and useful?

Answering that question is the business for this chapter. Chaotic systems, we will see, can be richly predictive in a variety of ways, even if sensitive dependence does place severe limits on the availability of *one* kind of prediction. This should perhaps already be clear, at least in outline: but it is well worth exploring the point in some detail.

4.2 Let's consider predictability in our now familiar paradigm, the Lorenz model. And for the moment, let's continue to focus on the predictability of individual trajectories: we will move on to consider a wider class of predictable dynamical properties in §4.5.

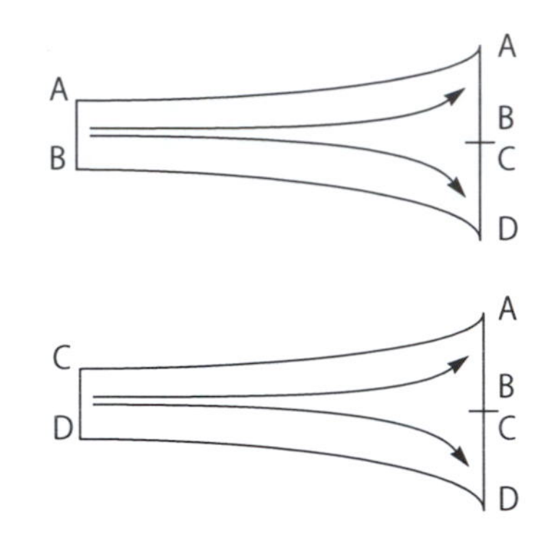

Figure 4.1 The structure L

In the Lorenz case, trajectories are pulled in very quickly towards the attractor; so what we need to understand is the behaviour of trajectories more or less on the attractor itself. To this end, it will be useful to consider first a simplification – in a phrase, we are briefly going to discuss a model of the Lorenz model. Take a structure L consisting of two equal sheets with the sheets of L then folded around to join each other again, as indicated by the labelling in Figure 4.1. Now imagine a possible toy dynamics defined on this structure such that the trajectories starting from any points on the baseline AD symmetrically spread apart. The resulting flow of trajectories is sketched in Figure 4.2 with the joins now made, and is obviously Lorenz-like in its basic geometry. It is a simple exemplar of a 'stretch and fold back' dynamics.

What happens to a trajectory passing through a point k on the unit line AD? We can specify k by means of an infinite binary expansion, as it might be .011010111... (measured from A). This point is in the segment AB; and so after one circuit on the first sheet – given the assumption of symmetric spreading – the trajectory crosses AD again at

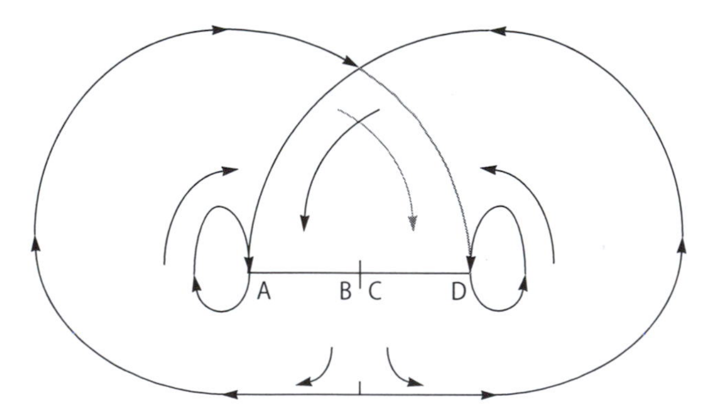

Figure 4.2 Lorenz-like dynamics on L

a point twice as far from A as k is. So the second crossing point k' is .11010111... (since a number in binary form is doubled by shifting its point one place to the right). This new crossing point is in the segment CD, and after another circuit (this time on the second sheet) the trajectory crosses AD again at a point whose distance from A is twice the distance that k' is from C, i.e. twice the difference between .1 and .11010111..., i.e. twice .01010111..., i.e. at the point .1010111.... And a little reflection shows that every further circuit as it were throws away another leading digit in the binary expansion of the crossing point. Generalizing, start at the point with binary expansion $.s_1s_2s_3s_4s_5...$ on the baseline; then successive crossing points will be $.s_2s_3s_4s_5s_6...$, $.s_3s_4s_5s_6s_7...$ and so on. (We can call this type of behaviour, for obvious reasons, a *symbol shift* dynamics.)

This way of looking at what happens on L immediately yields two initial facts about our toy dynamics. (1) If we know the first n digits in the binary expansion of the initial point k, then we can predict the next $n-1$ crossing points, though with one digit's less accuracy each time. So we can predict the order in which a trajectory through k visits the sheets of L for the next n circuits. However, such predictions are *exponentially expensive* in the initial data: we need to double the accuracy of our fix on k (i.e. discover another digit of its binary expansion) in order to predict the whereabouts of the trajectory for just one more circuit. (2) If we *only* know the first n binary digits of k, then what happens to the trajectory through k once n circuits have been completed will be quite unpredictable – we'll have lost e.g. even the most coarse-grained information about which sheet of L the trajectory is on. In other words, even if k and k^* agree on their first n digits, the later behaviour of the trajectories through these neighbouring points can be entirely different (so we have here a basic type of sensitive dependence on fine differences in initial conditions). In fact, a trajectory's sequence of later visits to the sheets of L will be as unpredictable for us as a sequence of coin-tosses. For this later behaviour depends on the string of binary digits in the expansion of k from the unknown $(n+1)$st place onwards, and that could be just any sequence of 1's and 0's at all.

Two further basic facts about the dynamics on L should also be noted. Suppose that k is a rational number. Then the binary expansion of k will (eventually) start repeating – for example, k might be $.111101\overline{010}$, where the overlining indicates infinite repetition. After four circuits, the crossing point k' is $.\overline{101}010$, and after another couple of circuits the crossing point is $.1\overline{010}$ which is k' again. Hence k leads to a *periodic* trajectory that (eventually) repeats itself. And generalizing, (3) there will be a countably infinite number of periodic trajectories on L – corresponding to the countable infinity of rational crossing points on the unit line AD. But (4) that still leaves an uncountable number of aperiodic trajectories on L, corresponding to the remaining uncountable infinity of non-rational real numbers in the unit interval. So almost all trajectories (as the mathematicians put it, 'all but a set of measure zero') are aperiodic.

4.3 Our model-of-the-Lorenz-model isn't accurate of course. It would be improved by adjusting the joins so that the points A and D don't get mapped exactly to themselves after a circuit. But the crucial shortcoming is that it allows distinct trajectories to merge as they cross the baseline (e.g. the trajectories starting at $.000\overline{10}$ and $.111\overline{10}$ cross after three cycles at the *same* point $.\overline{10}$). Merging is impossible in the true Lorenz system; and to allow room for trajectories to cycle round without ever intersecting or merging, the real Lorenz attractor must have a 'depth' lacking in L. (See the interlude on 'Stretching and folding' for an explanation of why we will expect the attractor in fact to consist in an infinite stack of sheets – like filo pastry rolled out and folded infinitely often – and thus to have a fractal structure.)

Still, the imagined dynamics on L in fact gives us a pretty good qualitative understanding of the stretch-and-fold behaviour of trajectories once they are pulled in near to the Lorenz attractor.

First, given a fix on an initial point, we will be able to predict the course of a trajectory at least for a brief time, albeit with rapidly reducing accuracy. And the better the initial fix, the longer we will be able to predict e.g. which wing of the attractor the trajectory is cycling around. In fact, the following general principle holds:

(P) $(\forall\delta)(\forall t)(\exists\epsilon)$(the state after interval t can be fixed within δ by fixing the initial state within ϵ),

where here and henceforth δ, t, $\epsilon > 0$. However, in the Lorenz system as in the toy case, the allowable error ϵ in determining the initial state (for a given accuracy of prediction δ) shrinks exponentially with time. We have approximately

(P_L) $(\forall\delta)(\forall t)$(to fix the state after interval t within δ, the initial state needs to be fixed within $\delta/e^{\lambda t}$), where $\lambda \approx 0.9$.

Which means that we have to more than double the accuracy of our fix on the initial state to extend a prediction of given accuracy for another tick of the clock (i.e. when t is incremented by 1).

It is tempting, perhaps, to gloss (P) as saying that the system is predictable in principle for as long as we want ('just make the initial error ϵ small enough ...'). However, care is needed. If we continue thinking of the Lorenz system purely mathematically – so that 'initial conditions' can be whatever we stipulate them to be, and 'prediction' just means computation given initial conditions – then the gloss is perhaps harmless enough. But matters are quite otherwise if we have an eye on applications. For example, to know which wing of the Lorenz attractor a trajectory is on after (say) a hundred ticks of the clock requires fixing the initial condition with an accuracy of something like forty significant decimal places. So if we were to try to use the Lorenz equations to predict the behaviour of some physical system for that long, then we would need initial data of literally incredible precision. To press again the argument of §3.1, it will make no physical sense to suppose that the relevant worldly quantities can have such a degree of precision. And over and above that point, quantum mechanical effects will in any case ensure that there is a limit on the accuracy of the useful measurement on *any* macro-quantity. But if it would be physically impossible to get enough initial data for longer-term predictions in applications to the real world, we certainly should not suggest that such predictions are always available 'in principle'. What could that now mean?

The exponential condition in (P_L) obtains because – as already indicated in §1.5 – we have the following:

$$\text{(EXP)} \quad |x(t) - y(t)| \approx |x(0) - y(0)|.e^{\lambda t},$$

where in the Lorenz case $\lambda \approx 0.9$. And this in turn implies e.g.

$$(\text{SDIC}_\text{L}) \quad (\forall\delta)(\exists y(0))(\exists t)(|x(0) - y(0)| < \delta \text{ and } |x(t) - y(t)| > l),$$

where l is the width of one of the wings of the Lorenz attractor. In plain terms: a pair of trajectories starting as close as you like can eventually end up on opposite wings of the attractor. Which gives us the sensitive dependency result parallel to (2) in the previous section – fix the initial conditions as best we can, we always eventually lose all knowledge of the whereabouts of the trajectory.

Finally, it is worth commenting that the other two noted features of our model-of-the-Lorenz-model also seem to carry over to the real Lorenz case. For the Lorenz attractor consists (it is believed) of a whole bunch of possible trajectories. There is a countable number of periodic ones; but these are tangled up with an uncountable number of aperiodic trajectories – and it is the preponderance of these which dictates the 'typical' behaviour of the system.

To summarize on prediction so far: However good our initial information, the detailed behaviour of a trajectory in the Lorenz system soon becomes quite unpredictable. But we *can* predict the behaviour of individual trajectories in the short term – for (P) holds. It is just that trajectory prediction is expensive in the initial data. Moreover, as has already been stressed in previous chapters, we can also predict the typical long-term shape and general location of trajectories – since we know they end up winding around the attractor, and we know about the structure of the attractor. So, despite sensitive dependence, the situation is certainly not predictively hopeless.

More on the continuity principle (P)

We have met principle (P) before. In the very first interlude in Chapter 1, 'Dynamical systems', we noted that if a dynamical system in canonical form meets a certain mild constraint, then its solutions will be 'continuous in the initial data'. And this was exactly the claim (P) that the distance $|x(t) - y(t)|$ can be made arbitrarily small by making $|x_0 - y_0|$ small enough. Hence (P), and such predictability as goes with it, holds of *all* 'nice' dynamical systems, including the chaotic ones. It is unfortunate, then, that one of the first papers on chaos written by a philosopher – Greg Hunt's 'Determinism, predictability and chaos' (Hunt, 1987) – claims, or seems to claim, that (P) is *not* true of chaotic systems. Since the error has propagated, we had better pause to set the record straight.

Hunt correctly notes that predictability requires continuity in the initial data; but he then writes

> A class of systems have been discovered whose behaviour ... fails the continuity condition in a radical way. These are the chaotic systems. No matter how close two systems are initially, their phase space paths may diverge arbitrarily far. Between any two points whose path ends later in, say, area A there will be a point whose path ends in B and *vice versa*. There is a dense intermingling of points whose subsequent paths become arbitrarily distant. ... There is no continuity between initial position and changes in final position. (Hunt 1987, 130–32)

So chaos, Hunt says, is flatly inconsistent with the thesis that 'predictions can be made arbitrarily accurate by making the determination of the initial state arbitrarily accurate'. Which certainly looks like a straight denial that (P) applies to chaotic systems. However, this is just a technical error. It is simply false that between any two points whose trajectory ends up on one wing of the Lorenz attractor at time *t*, there will be another point whose trajectory will end up on the other wing at *t*. Similarly for other standard examples of chaotic systems.

There's an oddity, too, in Hunt's talk about a 'final' position (if this means any more than 'position at some later time *t*') – trajectories spiral around a strange attractor like Lorenz's for ever, and hence there is no *final* position. The charitable interpretation, however, is that what Hunt actually has in mind is not ceaseless motion round a strange attractor, but (very differently) complex asymptotic motion towards a *simple* attractor.

For a vivid illustration of this kind of case, consider the desktop toy which consists of a pendulum suspended over three magnets arranged in a triangle. When the pendulum is released from an arbitrary starting point, it follows a more or less contorted and often apparently random path looping around the magnets, eventually more or less coming to rest hovering uneasily over one of them. We can model the idealized dynamics quite easily, with a well-behaved set of four interlocking differential equations. The attractors of the system are, as you would expect, just three points. Which is to say that, in the model, the state asymptotically approaches one or other of the points representing equilibrium states of rest, with the pendulum hanging over a magnet. But the possible asymptotic motions towards these point attractors are, at least in the mathematical model, unlimitedly complex.

Suppose we make a two-dimensional map of initial starting points for the pendulum bob, constructed as follows. Label the three magnets Red, Green and Blue: then colour a point *P* red if, when the pendulum is released at rest at point *P*, it ends up (according to the dynamical model) hovering over the Red magnet; similarly colour the starting point green or blue if the pendulum would end up hovering over the Green or Blue magnet. The total red area represents the basin of attraction of the attractor corresponding to the Red magnet, and so forth. It turns out that, in the model, these basins of attraction have wildly convoluted fractal boundaries. In some regions, wherever two basins seem to meet, we discover on closer examination (running the equations on a computer) that the third basin extends between them, and so apparently ad infinitum. In such a region, as Hunt almost says, between any two points whose path ends later at the Red magnet there will be a point whose path ends at the Green magnet and *vice versa*. And

(E) When initial states are within such a 'mixed' region, then we can not reduce error concerning the final location of the pendulum to any arbitrarily small quantity δ by suitably reducing the error in fixing the initial state to within some ϵ.

The potential final error will in fact always stay the same (namely, the distance between the point attractors at the magnets Red, Green and Blue).

So this case illustrates how there can be an intricate dependency relation between initial states and asymptotically approached final states. But we must be very careful in extracting from (E) any point about prediction. In fact, (E) is entirely consistent with the continuity principle (P); the model for the pendulum-over-magnets is still one in which we can notionally make predictive error after time t arbitrarily small.

How can (P) and (E) hold together? Well, to emphasize once more, the approach to the point attractor in the model is asymptotic, i.e. talk in (E) of the 'final' location of the pendulum means its position in the limit as *time goes to infinity* (remember, we *are* working here within an idealized mathematical model). So all the current case shows is that we can lose predictability-within-small-δ once we allow for indefinitely long time lapses. Thus, to be sure, the model doesn't satisfy the principle

(H) $(\forall\delta)(\exists\epsilon)(\forall t)$(the state after interval t can be fixed within δ by fixing the initial state within ϵ)

which reverses the inner two quantifiers in (P). But we hardly need to invoke fancy examples from modern dynamical systems theory to show that there are cases where (H) fails. It almost always *does* fail. Consider for instance a particle travelling on a straight line whose position at time t is given as $x_0.(1+t)$. This simple motion inflates any error in determining the initial position x_0 linearly with time, and even linear inflation is enough to defeat (H).

4.4 Chaotic dynamical models are not predictively useless. We can use them for short-term trajectory tracking. And we can extract long-term information about the general behaviour of trajectories as they wind round attractors. But there is an interesting complication about the latter claim which ought to be mentioned at this point.

Suppose we have a set of dynamical equations for $x_1, x_2, \ldots, x_n$ which don't have a nice solution. How do we actually extract any predictions from them? By numerical integration on a computer. Suppose initially we set $x_i = x_i(0)$: then in effect we calculate a sequence of steps $x_i(\delta)$, $x_i(2\delta)$, $x_i(3\delta)$ … until we reach values for the desired $x_i(t)$. At each step an error gets introduced – for numerical integration algorithms aren't perfect (and whether or not we are running a perfect algorithm, computers introduce round-off errors). But that raises a problem. If we think we have reason to believe that the abstract dynamical system in question has sensitive dependence on initial conditions, i.e. that errors exponentially inflate, what reason can we have for supposing that the error-ridden trajectories calculated by computer behave like the true trajectories of the abstract system?

The situation has a whiff of paradox. We may think that a system is sensitively dependent because of computer trials: but then, if it is sensitively dependent, why trust the computer trials? Likewise for other features that seem to show up computationally. Maybe the calculated trajectories of the Lorenz system, for example, wind around the two wings of the attractor hopping from one wing to the other in an apparently random way. But couldn't this very feature be an effect of the accumulation of random errors introduced by less-than-perfect computation? What warrants the inference that the true trajectories of the system visit wings of the trajectory in random sequence?

A good question, but one to which there are the makings of a good answer. There are *shadowing theorems* that provide the needed warrants at least in some relevant cases (and we might reasonably hope for generalizations). The content of such theorems is roughly as follows. If we try to work out what happens to the trajectory of some point $x(0)$, then our error-prone computation may soon diverge, and diverge exponentially, from the true trajectory. However, for well-behaved systems, the computed trajectory *will* approximate – will closely 'shadow' – the trajectory from some *other* nearby starting point $x'(0)$. Hence the behaviour of the computed trajectory will still give information about the behaviour of a trajectory of the system (just not the one we thought we were computing!).

In a little more detail, suppose that taking intervals along the true trajectory from $x(0)$ gives the steps

$$x(0) \Rightarrow x(\delta) \Rightarrow x(2\delta) \Rightarrow x(3\delta) \Rightarrow \ldots$$

while the noisy, error-prone computation from $\tilde{x}(0) = x(0)$ yields

$$\tilde{x}(0) \Rightarrow \tilde{x}(\delta) \Rightarrow \tilde{x}(2\delta) \Rightarrow \tilde{x}(3\delta) \Rightarrow \ldots.$$

Then the error $|x(n\delta) - \tilde{x}(n\delta)|$ may indeed tend to grow exponentially with n, so that the computed trajectory very quickly peels away from the true one. However, under certain conditions applying to most dynamical systems which contract phase space volumes (see §1.5), there will be *another* starting point $x'(0)$, initiating a trajectory with the time-evolution

$$x'(0) \Rightarrow x'(\delta) \Rightarrow x'(2\delta) \Rightarrow x'(3\delta) \Rightarrow \ldots$$

such that for all n, $|x'(n\delta) - \tilde{x}(n\delta)| < \epsilon$, where the constant error term ϵ depends in part on the size of the computational error introduced at each step from $\tilde{x}(0)$. Hence the computed trajectory from $\tilde{x}(0)$ shadows within ϵ the true trajectory starting from somewhere else nearby. And that's good enough, at least when we are interested in using long-term computations to extract qualitative information about the general sorts of things that happen to trajectories in the system under examination.

Moral: shadowing theorems – where available – warrant innocent reliance on computers.

4.5 So far, we have repeatedly considered the Lorenz equations for Rayleigh-Bénard flow (§1.4) with a certain fixed setting of the parameters. It's time to generalize a little. So let's now take the equations in the slightly more general form:

$$(L^*) \quad \begin{aligned} dx/dt &= 10(y - x) \\ dy/dt &= rx - y - xz \\ dz/dt &= xy - 8z/3. \end{aligned}$$

Here we have still kept fixed the numerical values of the parameters which appear in the first and third equations; but the parameter in the second equation – which was earlier set at 28 – has now been replaced by the adjustable r. Exploration by computer rapidly reveals that for different values of r the solutions for (L^*) can behave in markedly different ways.

What, given the derivation of the Lorenz equations, is the physical significance of the parameter r? The full story doesn't matter; the headline point is that r is directly proportional to the constant temperature difference δT between the top and bottom of the 'box' containing the flow (compare Figure 1.4). The setting for δT is of course easy to manipulate in experimental contexts. Thus we can readily test to find how the overall character of the convection actually changes with the value of the parameter r.

In short, the generalized Lorenz model delivers a raft of predictions about the correlation between values of r and patterns of dynamical behaviour – predictions which can be experimentally checked.

To explore further, let's note a few salient facts. At $x = y = z = 0$, the derivatives in (L^*) are all zero. So the steady state where the system stays put at the phase space origin $(0, 0, 0)$ is always a solution: this corresponds to a state of the fluid with no convection at all. If $0 < r < 1$, this steady state is stable and indeed a global attractor, meaning that all trajectories are pulled into the origin. In physical terms, the model says that small temperature differences between top and bottom of the box will not be enough to maintain convection, so that any initial circulatory motion eventually dies away.

When r passes through 1, however, the origin becomes unstable: in other words, most trajectories starting near the origin now move *away* from it – i.e. the slightest nudge to the no-convection state typically starts the fluid circulating. But the origin gives birth, so to speak, to a pair of new attracting states: i.e. there is now a pair of phase space points C^- and C^+ (representing steady states of convection, either clockwise or anticlockwise) such that all trajectories wind in towards one or other of these points. As r increases, the points C^- and C^+ (symmetrically placed on either side of the origin) get further apart. In physical terms, this means that when the temperature difference δT is large enough the system settles down to steady convection in one direction or the other, with the circulation velocity increasing as δT increases.

When r increases further past the critical value $r_H \approx 24.74$, the points C^- and C^+ in turn become unstable (trajectories starting nearby wind now away from them) but a strange attractor of the now familiar type is found. (In fact, for $r > r_H$ but not too large, as with our original setting of $r = 28$, the points C^+ and C^- correspond to the 'holes' in the middle of the wings of Lorenz-like attractors.)

So, for small values of r, the equations tell us that motion gets damped away, for larger values steady convection is predicted, for larger values again the prediction is a chaotic regime with fluttering, randomly reversing, convection. And these qualitative predictions come with considerable quantitative details. We get predictions of the critical values of δT where one regime is replaced by another. We get predictions e.g. of the way that circulation velocity depends on δT in the pre-chaotic regime $1 < r < r_H$. And in the chaotic regime $r > r_H$, there will be estimates of the size and fractal dimension of the strange attractor for different r, and of various quantities associated with motion near the attractor (such as the relative speeds at which trajectory bundles are shrunk in one direction and spread in another – see again Figure 3.2a).

Now, this is only the very beginning of a fascinatingly complex story; but it already illustrates a theme which is absolutely central to modern dynamical theory. For what we see here is how equations like (L^*) generate very detailed correlations between *values of controllable parameters* and qualitative and quantitative *features of the dynamics*. There is nothing novel about the idea of trying to find such correlations: the classical theory of the simple pendulum – to take the most humdrum example – predicts a certain correlation between pendulum length (parameter) and the period of swing (dynamical feature). This prediction is if anything more interesting, and certainly is easier to test, than e.g. predictions of the exact pendulum trajectory for a certain setting of the parameter and initial state. Likewise in cases like Lorenz's, except now the parameter/dynamics correlations can be vastly more complex than anything encountered in high-school mechanics. Such predicted correlations are of the greatest interest – and they can often be more or less readily tested.

In sum, once we look for more than long-term stories about particular trajectories, we see that chaotic models – i.e. models like (L*) which have chaotic regimes for at least some parameter settings – can be *richly* predictive.

Bifurcations in the Lorenz system

The full story about the possible behaviours of solutions of the Lorenz equations (L*) for different values of r is startlingly convoluted. We will here say just a little more about the behaviour of solutions for $r < r_H$, and also something about some large r solutions (the discussion will inevitably remain extremely sketchy). The aim is to give some feeling for the complexity of the correlations between parameter-settings and dynamics, and to give an initial sense of some ways in which chaos can arise.

Start with the simple observation that the differentials on the left of (L*) will be zero when $y = x$ (from the first equation), $z = 3x^2/8$ (from the third), and so – putting these equalities into the middle equation – when

$$x^3 - \beta x = 0, \qquad \text{where } \beta = 8(r-1)/3.$$

If $r < 1$, β is negative, and so the origin $x = y = z = 0$ is the only 'stationary point' where the differentials vanish (yielding a steady state solution with the variables taking a constant value over time). But for $r \geq 1$ the differentials also vanish at

$$C^- = (-\sqrt{\beta}, -\sqrt{\beta}, 3\beta/8)$$
$$C^+ = (+\sqrt{\beta}, +\sqrt{\beta}, 3\beta/8).$$

These two points coincide with the origin when $r = 1$, but then there is a *bifurcation*, giving birth to two new stationary points which drift away from the origin as r increases: these points (to repeat) represent states of steady convection at constant velocity in one direction or the other,

There are standard techniques for analysing behaviour in the neighbourhood of stationary points for systems of first-order differential equations like (L*). They show that when $r < 1$, the origin is an attractor: but for $r > 1$, although the origin still pulls in those trajectories lying in a certain sheet M through the origin, it repels all other trajectories away, which get attracted instead to either C^- or C^+ (see Figure 4.3a, which indicates what happens to trajectories starting near the origin). Initially, for smaller r, trajectories repelled away from the origin simply approach the nearer of the two attractors C^- and C^+. But as r increases, trajectories start to spiral round the attractors in looser

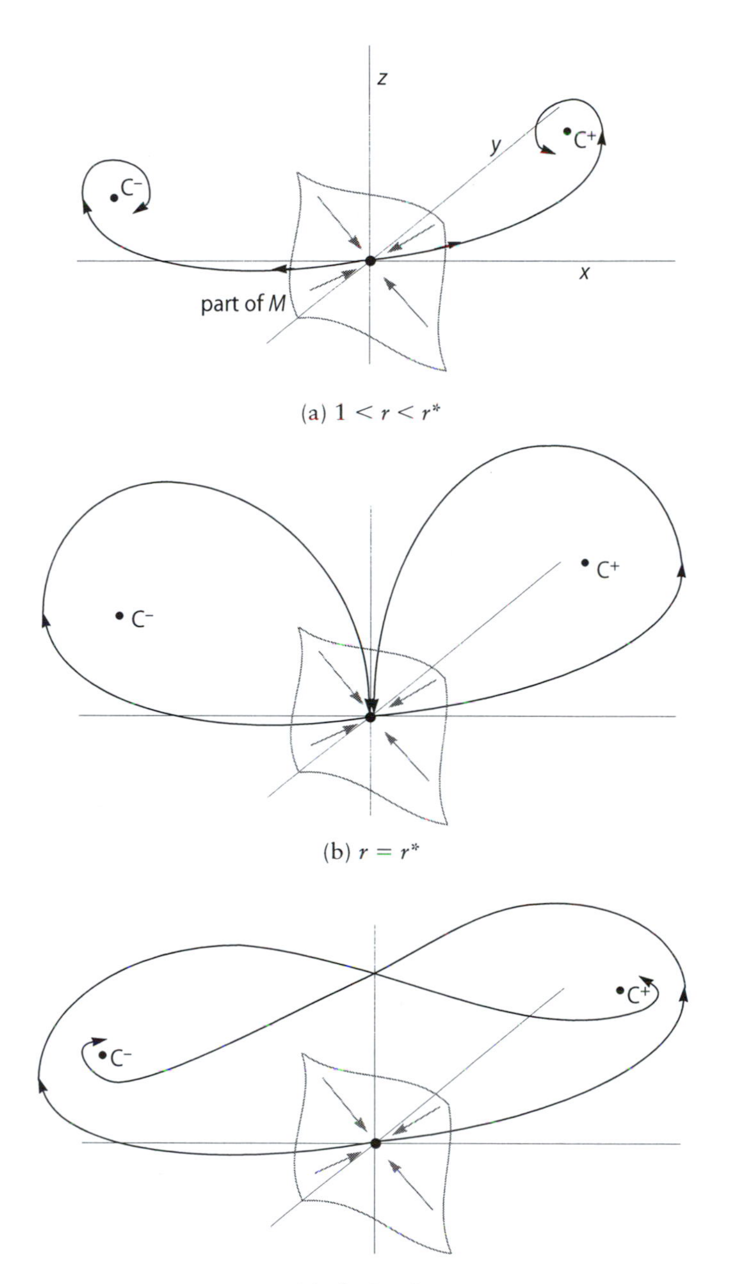

Figure 4.3 Behaviour of trajectories starting near the origin

and looser loops, until the critical value $r^* \approx 13.93$ where a trajectory repelled from the vicinity of the origin circles right back out to the origin itself – a trajectory like this which tends to the same point as time goes both to $-\infty$ and $+\infty$ is called *homoclinic*. (See Figure 4.3b: the full journey would indeed take infinitely long, as the trajectory gets ever slower as it gets nearer the origin). For r above the critical value r^*, trajectories from near the origin start 'crossing over' to the opposite side from the origin, and are then captured by the other attractor, as in Figure 4.3c. The result is a flow of trajectories into the two attractors a bit like an imploding Lorenz attractor (Figure 1.3), with trajectories falling into the 'holes' in the centres of the two wings. Then, when r increases past $r_H \approx 24.74$, the two points C^- and C^+ cease to attract, and the flow of trajectories winding round them no longer implodes but keeps looping round for ever – and we are into the familiar Lorenz regime.

Consider next what happens for some larger values of r, where chaos dies away again (only to return for yet larger values of r). If we set $r = 100$, and calculate trajectories on a computer, we find that the Lorenz attractor has now disappeared, and all numerically calculated trajectories seem to be attracted instead to one or other of two very simple periodic orbits. One of these attracting trajectories does a double loop round C^-, swings out to loop loosely once round C^+, and then rejoins itself; its mirror image loops twice round C^+ and once round C^-.

Where's the chaos gone? Or to put it the other way around, how does a Lorenz style chaotic attractor re-emerge as we reduce the value of r back from 100? Well, we find that when we decrease r past about 99.98, each of those periodic orbits gets (as it were) slightly jolted – so that the first orbit, for example, after its two loops round C^- and one round C^+, doesn't *quite* repeat itself, but does another two loops/one loop circuit before getting back to its starting point (similarly for the mirror image: see the schematic Figure 4.4 for a sketch of this *period-doubling bifurcation* of one of the two orbits from a three-loop to a six-loop structure). As r is decreased past about 99.63, each of the two six-looped attractors splits again: there is another period-doubling, so now the attracting orbits do *four* two loop/one loop cycles before repeating themselves. And as r is reduced further, we quickly encounter further period-doublings – the two attractors for the system stay periodic, but with longer and longer periods. These bifurcations arrive ever faster as we reduce r. Analysis strongly suggests that there will be an infinite number of such period-doublings completed very quickly – in fact before r reaches 99.5. So by then, the finite periodicity of each of the symmetric pair of attractors is destroyed. Hence we must now be dealing with a more complex symmetric structure comprising (at least) some infinite trajectories. And in fact, by this *period-doubling route to chaos*, we are back with a Lorenz-like structure.

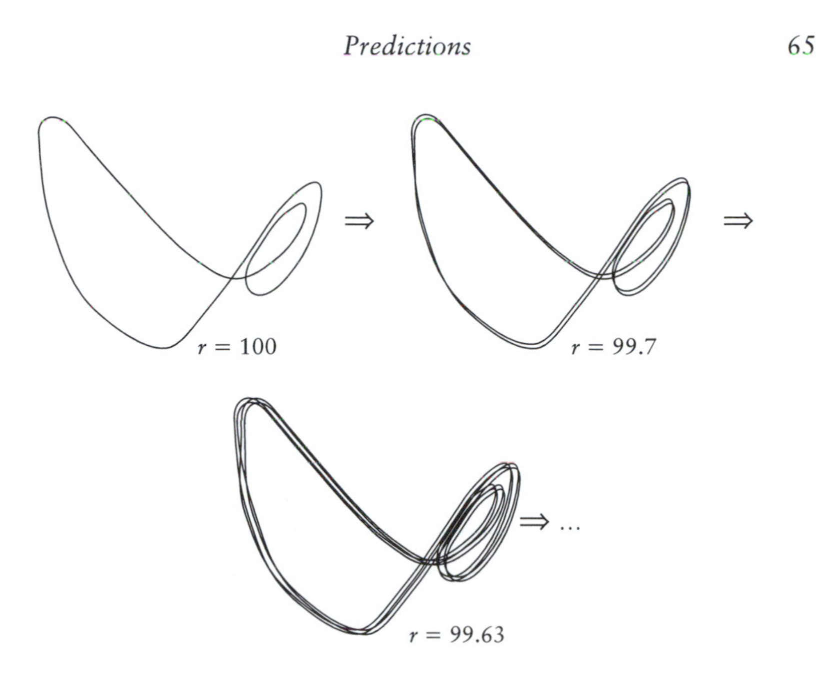

Figure 4.4 Period-doubling below $r = 100$

Let's briefly take stock. We've found chaos when *r* increases far enough past the value r^* when the orbit is homoclinic. And that's one way that chaos can get born, in the break-up of a homoclinic orbit (more about this in the final interlude of this chapter). And now we've sketched a period-doubling route to chaos as *r* decreases from $r = 100$ (more about this in Chapter 6). And these are only the beginnings of the complexities of the behaviour of the Lorenz model governed by (L*) for varying *r*.

Which may all raise a question: in the previous chapter, it was suggested that chaotic models, despite their intricacy, may exhibit a kind of intrinsic simplicity that makes them still potentially apt for use in applied models. But now it is beginning to seem that the Lorenz model is quite wildly complex after all. Was the perceived 'simplicity' merely a temporary illusion? Well again, the appearance of complexity is in some ways an artefact of the level at which we are telling the story. Think of things this way: for a particular value of *r*, there is a corresponding three-dimensional flow of possible trajectories – so what we really need to envisage (four dimensionally!) is how these three-dimensional structures get continuously bent and distorted with changing *r*. And just as simple stretches and folds can produce fractal monsters such as Lorenz attractors, quite simple distortions-as-*r*-changes are in fact producing the very complex effects noted here. In other words, with some effort, the equation-level simplicity in just varying the single parameter *r* can indeed be seen as correlated with an abstract structural simplicity. But to give a detailed sense of that more abstract story would take us too far afield (though the final interlude of this chapter gives some more ingredients for the story).

4.6 At the beginning of this chapter we posed the question: if chaotic models give badly wrong predictions when applied to the world (because of their sensitive dependence on initial conditions), how come that they are *models*? We have now spelt out the obvious answer in some detail. True enough, applied chaotic models can't be used to predict the detailed time-evolution of the relevant physical quantities over long periods. But they can deliver a range of *other* kinds of prediction, both qualitative and quantitative, about how a system works for given parameter settings. And (often more importantly) they also can yield qualitative and quantitative predictions concerning the way the behaviour of the system changes with changing settings of the relevant parameters. The Lorenz model, for example, delivers a package of predictions about the character of Rayleigh-Bénard flow at various temperature differences δT.

Now, these latter predictions of the Lorenz model turn out to be mostly *false*, especially for large *r* regimes (hardly surprisingly, perhaps, given the radical simplifications that went into the model's construction in the first place). But given our current purposes, the fact that our paradigm is not a terrific empirical success, at least in its initially intended application, is of no great moment – though it would of course be bad news if, counterfactually, there were *no* even partial empirical successes for chaotic models. Our concerns here are mostly methodological: we want to understand the *sort* of models involved in chaos theory, the *kind* of explanations they might deliver (despite their empirically unreal fractal intricacy) and so on, and the Lorenz model remains a useful test exemplar.

Still, the fact that this model is, relatively speaking, an empirical failure should at least give pause e.g. to casual extrapolation to talk about a 'Butterfly Effect' – a point that is worth brief comment. Noting that the Lorenz model is intended to represent atmospheric convection, we are supposed to leap to conclusions about the way that the weather system itself, on a large-scale level, is affected by arbitrarily small changes. For example, Ian Stewart (Stewart 1989, 141) tells us that

> The flapping of a butterfly's wing today produces a tiny change in the state of the atmosphere. Over a period of time, what the

atmosphere actually does diverges from what it would have done. So, in a month's time, a tornado that would have devastated the Indonesian coast doesn't happen. Or maybe one that wasn't going to happen, does.

But hold on! Even if we ignore for the moment the empirical shortcomings of the Lorenz model, how on earth are *tornadoes* supposed to get into the story? The model was intended to describe the behaviour inside one of a series of parallel horizontal convection rolls: and it actually counts *against* butterfly-sized causes producing tornado-like effects. For the model assumes that the large-scale pattern of rolls, laid side by side like so many felled logs, remains entirely stable: the chaotic behaviour is local, as the rolls change their rotation-speeds in never repeating patterns. A little nudge to the left-most roll will result, let's suppose, in a chaotically unpredictable change in the convection pattern of this roll. And let's suppose that an augmented model takes into account boundary effects between rolls in a way that allows such changes to propagate down the series of rolls. So we will get effectively unpredictable changes in the convection pattern in the right-most roll – but, so long as we are still working within the Lorenz paradigm, there is no destructive break-up of the rolls, no wildly accelerating convection, and hence certainly no tornadoes!

So there is a double problem in projecting from Lorenz to a real-world Butterfly Effect. First, the basic dynamical model in fact isn't that successful in coping even with the phenomenon of Rayleigh-Bénard flow. And second, if we pretend that the Lorenz model is better than it is in its target domain, it still doesn't entail anything about tornado-sized effects. The line of thought in defence of Butterfly Effects must be very much more speculative: e.g. 'Since even the very simple Lorenz model is chaotic, and moreover – now we know what to look for – we find that chaotic regimes are rather pervasive in non-linear systems, it is reasonable to conjecture that *any* decent model of the large-scale behaviour of the atmosphere will also exhibit the sort of exponential sensitivity to initial conditions that can make butterfly-sized perturbations determine the development of tornados.' But note this *is* just a speculation; I suspect that no-one yet knows how to begin assessing the rational odds for betting on its truth.

Homoclinic explosions

This concluding excursus is very much by way of an appendix, and can readily be skipped. But the mathematically inclined may find it illuminating to consider in just a little more detail how the break-up of a homoclinic orbit in a Lorenz-like system can give birth to chaos, and we will see again how a quite simple operation on trajectory bundles can give rise to a fractal structure. (Readers interested in the nature of explanations in mathematics might also like to reflect on the kind of understanding engendered by the sort of quick and dirty presentation which follows.)

Start again from the situation sketched in Figure 4.3b, and we will consider what happens near the origin when r *just* exceeds the critical value r^*, when trajectories are not quite homoclinic. Enclose the origin in a little box, and take a pair of bundles of trajectories emanating from the box, on opposite sides of M (M, recall, is the sheet of trajectories still pulled in to the origin). Nothing very exciting happens to these trajectory bundles on their almost homoclinic loops round respectively C^- and C^+ – i.e. the bundles stay in one

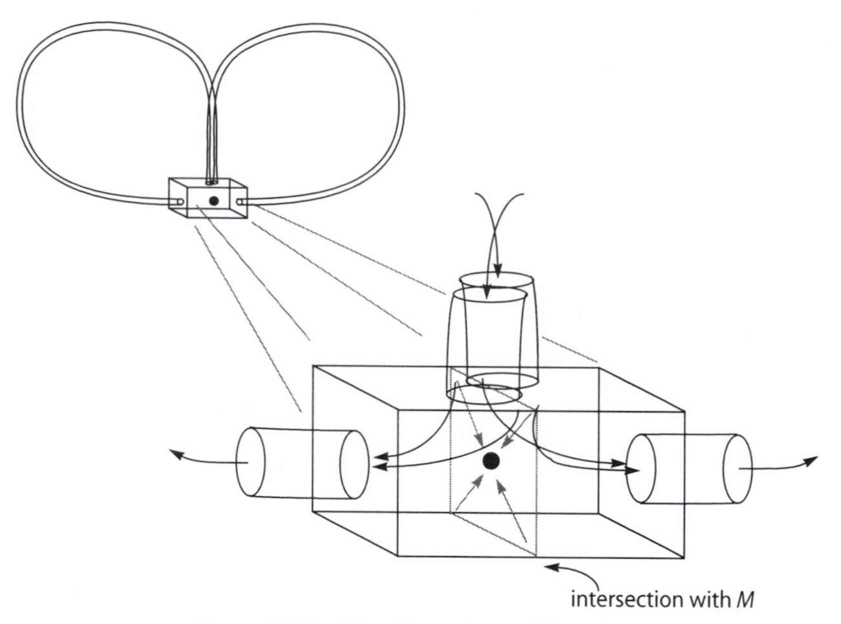

Figure 4.5 Bundles of near-homoclinic orbits

piece for that part of the circuit, and all the action happens when they revisit the neighbourhood of origin and enter the box again (Figure 4.5). Trajectories that hit the top of the box near the intersection with M will first be squeezed together very sharply as they *almost* visit the origin together before being repelled away; trajectories that hit the top of the box further from M will be less affected. To see what this means, imagine a rectangle *ABCD* on the top of the box, straddling the intersection with M (Figure 4.6).

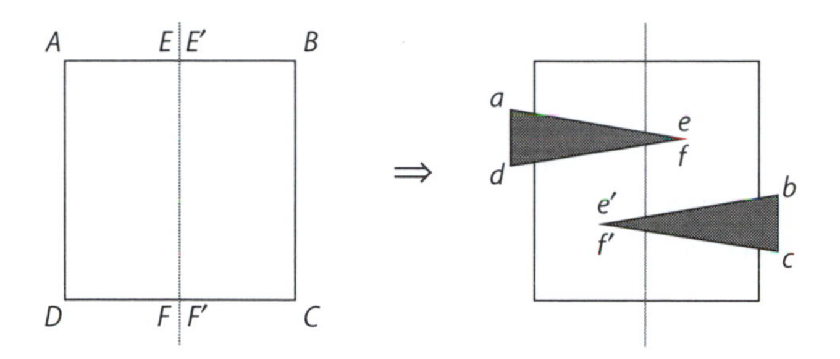

Figure 4.6 The 'copying' effect of near-homoclinic orbits

The bundle of trajectories through points in the left half *AEFD* (where *EF* is arbitrarily close to *M* from the left) gets compressed by the dynamics inside the box into a wedge shape – since trajectories through *EF* almost converge to a point before being thrown out of the box, while trajectories through AB are squeezed more mildly. This wedge of trajectories is then thrown out of the box to the left, to cycle around with a fairly constant profile to hit the top of the box again in the wedge *aefd* – with the point of the wedge in the right half of the initial rectangle (since we are considering the case where r is slightly greater than r^*, trajectories emanating very near the origin slightly overshoot the origin on their return). Symmetrically for trajectories starting from *BE′F′C*.

In sum: If r is just past the critical value for a homoclinic orbit, the effect of the dynamics is to 'copy' points in the rectangle *ABCD* after one circuit into a couple of wedges overlapping the rectangle. But this is obviously exactly the same sort of effect as is produced by a Barnsley copier (see the interlude in §2.4 – although previously we had the 'lenses' producing copies of the whole original, whereas here we have image-splitting). So we will naturally expect the set of points Q produced by 'copying' the rectangle *ABCD* in the manner of Figure 4.6 an unlimited number of times to be a fractal which is invariant under further copying: which indeed it is (it's a Cantor-like gappy fractal). And what is the significance of Q? Well, by definition, a trajectory starting from a point in Q always returns to hit the rectangle *ABCD* within Q again: so Q is in fact the set of crossing points in *ABCD* for those trajectories – call them Q-trajectories – which always return to the neighbourhood of the origin, never getting captured by C^- or C^+.

What the last two paragraphs indicate therefore is this: when r increases past the critical value r^*, the homoclinic trajectory is destroyed – leaving as debris a whole bunch of Q-trajectories which still cycle round in an almost homoclinic way, always returning to the neighbourhood of the origin. And this bunch of trajectories has a fractal geometry, intersecting the rectangle *ABCD* in the Cantor-like set *Q*. So the breakdown of the homoclinic trajectory has suddenly given birth to a strange two-looped fractal structure. This structure isn't yet an attractor – almost all other trajectories will quickly get trapped by either C^- or C^+ – but we can see in it the seed of the Lorenz attractor. And as r increases, this structure grows and imposes more influence until $r \approx 24.06$ when it becomes an attractor; then, only slightly later (at $r = r_H \approx 24.74$) C^- and C^+ cease to attract, and the strange attractor uniquely holds the field – we are at last in the familiar regime of Lorenz chaos.

So that's one remarkable way that a fractal attractor can be born, in a *homoclinic explosion*.

5

Approximate truth

5.1 In Chapter 3, it was argued that worldly phenomena of the kinds typically modelled by chaotic theories cannot exemplify in their time-evolutions the infinitely intricate patterns characteristic of chaos. So even if (as argued in Chapter 4) chaotic theories can be richly predictive, it seems that they cannot be strictly *true*.

Is this a kind of scepticism about chaos? Not specifically. Dynamical models in classical macrophysics always postulate an infinite precision in the values of the relevant quantities, yet we typically have excellent physical reasons to suppose that these quantities cannot take infinitely precise values. So it is not only chaotic theories that idealize: most of macrophysics is in just the same boat. Does a more global scepticism threaten, then? Surely not. The natural line – implicit, I think, in the comments of working scientists if and when they address the issue – is to allow that such macrophysical theories may idealize, may fail to be true if interpreted by perhaps over-strict standards, but to insist that they can still be more or less *approximately* true (and can be *known* to be approximately true).

But what does approximate truth amount to? For reasons that will emerge, it is doubtful that there is any story to be told which is both substantive and general. So let's continue to concentrate on a restricted class of cases – classical dynamical theories that can be regimented by means of a system of linked first-order differential equations of form (D) (§1.1). A mathematical modelling of this kind, as we have seen, specifies a certain geometric structure, a bundle of allowed trajectories in an appropriate phase space. (More generally, there is a family of such bundles, one bundle for each different setting for the values of the parameters in the equations.) And when such a dynamical structure is put to work to model some physical system, how does it match up to the world? The set of physically possible behaviours of the modelled dynamical system (for a given setting of the control parameters) can also be represented as constituting a geometric structure (§3.2). For every physically possible behaviour, there will be a corresponding time-evolution of the relevant physical quantities. If we pretend for a moment that the values of these physical quantities are perfectly determinate, then we can encode the values at a given time by a point in a suitable phase space; as values evolve from some initial state, the representing point can then be thought of as tracing out a phase space trajectory. So there will be a bundle of such (abstract) trajectories to represent the physically possible time-evolutions of the real-world dynamical system. Which gives us two geometric structures – one doing the modelling, the other encoding what needs to be modelled. If these are replicas, then we can say that the dynamical theory that postulates such a model is true, period. And if the structures are similar enough, we can say that the dynamical theory in question is approximately true.

Note, a dynamical model of the kind that we are interested in tells us *how* the state variables evolve over time (by telling us how their rates of change relate to current values). But it doesn't in itself tell us *why* the variables evolve as they do, in the sense of explicitly giving a causal-explanatory story. To be sure, we may build a dynamical theory on the basis of an understanding of what is going on causally – e.g. by putting terms on the right-hand side of our equations to represent various causal factors affecting the time evolution of the relevant quantities. However, while causal hypotheses may motivate our equations, they are no part of the content of the resultant dynamical model. That only tells us what would happen over time from various possible initial states: and it will be (approximately) correct just so long as that *is* what would happen – i.e. if the phase space trajectories in the model (approximately) match the trajectories representing actual worldly possibilities.

Take a simple example, the textbook account of the dynamics of a freely swinging pendulum. This characterizes a pure abstraction, the ideal frictionless pendulum moving in a plane according to Newton's laws and subject only to a constant vertical force. The governing equations determine the allowable patterns for the time-evolution of the ideal pendulum's angular displacement and velocity as a function of the pendulum's fixed length etc. If we conceive of plotting a three-dimensional graph of time against displacement against velocity, then a certain bundle of three-dimensional curves will trace the allowable behaviours of an ideal pendulum of given length subject to a given force. If we conceive, yet more abstractly, of these three-dimensional bundles being 'plotted' against pendulum length and applied force, we will get a more complex five-dimensional structure that in addition encodes the way that the possible behaviours of the pendulum depend on the length and force. This abstract structure gets put to empirical work, however, when it is claimed – as a first shot – that it exactly matches a structure that similarly would encode the physically possible behaviours of real freely swinging pendulums of varying lengths, etc.

Of course, this latter empirical claim is false. For a start, real physical pendulums are damped by friction, air-resistance, and so on, and eventually stop moving: ideal pendulums swing for ever. Still, considered for shortish periods, real pendulums behave similarly to ideal pendulums. In more geometrical terms, a medium-thick time-slice of the mathematically defined abstract structure will give, curve by curve in the bundle, a decent approximation to a time-slice of the corresponding structure encoding the possible behaviour of real pendulums. In that sense, the textbook treatment of the pendulum counts as approximately true (or approximately true, considered for short-enough periods).

Generalizing, we can say that a dynamical theory is approximately true just if the modelling geometric structure approximates (in suitable respects) to the structure to be modelled: a basic case is where trajectories in the model closely track trajectories encoding physically real behaviours (or at least, track them for long enough). True, since the relevant physical quantities are in fact typically coarse-grained, the phase space trajectories which represent their time evolution should really be fuzzily drawn. But no matter. The same line on approximate truth should still apply (trajectories in a precise model can approximately track fuzzy paths). This complication apart, there remains the problem of spelling out formal approximation relations between trajectory bundles to capture various intuitive relations such as 'close-tracking'. But getting the fine details to work now looks to be a set of problems in geometry – there's no residual conceptual issue about the very idea of approximate truth.

Call this the *naïve account* of approximate truth for applied dynamical models. This chapter argues that the naïve account is the right sort of account. I suspect that few dynamical theorists have ever supposed otherwise: however, there is a tendency in many philosophical discussions of approximate truth to find the very notion deeply problematic. So, at least for the philosophical reader troubled by such discussions, I need to show that the naïve account is indeed in good order (or at least, that it is in good enough shape for our purposes). Non-philosophers who are happy in their naïvety, or those who want to press on to get to know more about chaos, may cheerfully jump to Chapter 6.

5.2 I claim that

[A] 'P' is approximately true if and only if approximately P.

And furthermore,

[Exp] The order of explanation goes from right to left across the biconditional [A].

Hence the task of construing the notion of approximate truth reduces to the task of explaining how the modifier 'approximately' gets applied to propositions of various types.

[A] is an instance of a more general principle: with minor explicable exceptions, we have

[M] 'P' is *M*-ly true if and only if *M*-ly P,

(for adverbs *M*-ly such that '*M*-ly true' makes sense), with the right-hand side of the biconditional again explaining the content of the left. There is a little more about this general line in the following interlude; for the moment, however, let's just note that it should appeal to anyone persuaded by a broadly deflationist account of the way the truth-predicate works within the class of truth-apt statements. To take the simplest deflationist story: suppose that 'is true' is simply used to endorse assertions, perhaps without explicitly repeating them. The truth-predicate then has a merely formal *disquotational* role; '"P" is true' says no more than plain 'P'. Which means that the truth-predicate does not express a property that can come in various modes or degrees. Hence apparent modifications of the truth-predicate had better be understood as modifications of the proposition being said to be true. Thus just as '"P" is true' says no more than 'P', '"P" is allegedly/unfortunately/unexpectedly/probably true' says no more than 'Allegedly/unfortunately/unexpectedly/probably P'. Conversely, robust exceptions to [M] would sink disquotationalism by revealing truth as a more-than-formal property that can – as it were – come in its own proprietary flavours.

In particular, then, a disquotationalist about truth should endorse the application of [M] to 'approximately', i.e. should endorse the combination of [A] plus [Exp]. Likewise for other, less extreme, deflationists (see the following interlude). And indeed, metaphysically inflationary theorists about truth can also agree on this point about approximate truth. For suppose that you conceive of truth as a matter of the existence of some corresponding situation or fact (where you take a chunkily realistic view of facts). Existence does not admit of degrees; hence, in the last analysis, neither can truth. From this perspective, any account that aims to give content to some notion of approximate truth must respect the constraint that truth, like existence, is fundamentally an all-or-nothing matter. And then if truth is basically all-or-nothing, it is again attractive to explain apparent modifications of the truth-predicate by appeal once more to [A], read right-to-left.

Take a couple of very simple instances of [A]. (1) 'Jones is six foot tall' is approximately true because Jones is approximately six foot (henceforth, take 'approximately' to mean *at least* approximately – i.e. not to rule out strict truth). Here, in its second occurrence, 'approximately' serves to fuzzify the claim that Jones is six foot tall (or perhaps *further* fuzzify it, if it is thought that the vernacular claim is already somewhat vague): the degree of fuzzification indicated will be context-dependent. True, there are deep problems about the semantics of vague concepts, and hence serious problems about the semantics of a fuzzifying operator: but the thought is that there is no *additional* problem about understanding what it is for some proposition to be approximately true when this amounts to the truth of a corresponding fuzzified proposition.

Case (2): 'Snoopy is a spaniel' is approximately true because Snoopy is a spaniel, approximately enough at least for some contexts. Here, there isn't the same sort of vagueness: approximately being of one (more or less sharply-bounded) kind is naturally construed as a matter of belonging to some other (sharply-bounded) kind placed close enough on a hierarchical kinship tree. Thus it is close to the truth that Snoopy is a spaniel, while not approximately true that Snoopy is a bear, because spaniels are close enough to beagles on the zoological family tree, and bears are too far removed. (Some hold that it is odd to say that 'Snoopy is a spaniel' is approximately true; but such objectors seem to find it *equally* odd to say that Snoopy is approximately a spaniel – and that pair of reactions *confirms*, rather than subverts, the thesis that [A] holds. For that thesis implies that judgements about the degrees of acceptability of each side of the biconditional should march in step.)

Now, if the modifier 'approximately' is sensibly to be applied to a given proposition, then typically some focal term(s) in the proposition should be locatable in a domain which is subject to some natural ordering (as e.g. the class of *heights*, or of *mammalian kinds*). However, the detailed structure of the relevant domain for comparison, and what counts for closeness therein, can vary widely between types of proposition. But according to [Exp], applications of the idea of approximate truth are to be explained via the application of the approximation operator to whatever proposition is in question. So – as the examples (1) and (2) already indicate – different applications of the notion of approximate truth may well require different kinds of detailed elucidation.

Which gives us a principled reason why we shouldn't necessarily expect a substantive account of approximate truth for one domain to carry over to another. In particular, then, there is no requirement that a detailed account of what makes for the approximate truth of mathematically framed dynamical theories in the physical sciences should carry over readily to other cases of approximate truth, or vice versa.

But then, how *do* [A] and [Exp] bear on the case of dynamical theories?

Three comments on [M], [A] and [Exp]

Before proceeding with the main business, three quick comments.

First, what kinds of exceptions are there to [M] and [A]? [M] doesn't seem to apply e.g. to

'P' is absolutely/perfectly/very true.

nor to

'P' is partly/nearly all/wholly true.

But no matter: in the first kind of case, the adverb serves merely for emphasis; in the second kind of case, the forms are appropriate only when 'P' is suitably complex, and then they are respectively equivalent to

Part of/nearly all of/the whole of 'P' is true.

In neither sort of case do the adverbs descriptively modify the formal truth-predicate in the sort of way that a disquotationalist predicts is impossible ('partly' simply restricts the scope of its application).

The following sort of objection to [A] is sometimes suggested. The modern Mrs Malaprop's statement 'The milkman is erotic' may be approximately true since it approximates to the true claim 'The milkman is *erratic*' got by a small word-shift – even if it isn't the case, even approximately, that the man is erotic. However, the confusion here is evident: it is one thing to say that a sentence almost gets it right about how things are in the world, another thing to say that a sentence is very similar to a *different* sentence which does get things right. In a phrase, we must distinguish being close to the truth from being close to a truth: [A] applies to the former kind of approximation, not the latter.

There are objections in a not unrelated vein to other instances of [M]. Consider the modal case:

[P] 'P' is possibly true if and only if possibly P.

It may be claimed, for example, that the sentence 'All bachelors are married' could possibly have expressed a truth, once the possibility of suitable changes of meaning is in play, although it is not possible that all bachelors are married. (If we type words by their meanings, this is again a case involving word-shift.) But even the disquotationalist defender of [M] can live with this special exception. If we semantically ascend from talk of bachelors to talk of the word 'bachelors', but then allow meaning changes (word-shifts), it should occasion no surprise that disquotational descent won't take us back to our starting-point.

Second, we argued that [M], [A] and [Exp] must appeal to the pure disquotationalist. But in fact, as already indicated, the same claims should seem more or less equally compelling across the spectrum of deflationary theories. Consider just one variant. Pure disquotationalism holds that *any* sentence which is syntactically apt to be substituted in schema

[D] 'P' is true if and only if P,

and is governed by canons of correct use can properly be attributed truth or falsehood. Some philosophers want to be rather more discriminating: perhaps some such sentences are not truth-apt in virtue of not expressing genuine beliefs (here a substantial philosophy of mind will be invoked, as when a Humean argues e.g. that evaluative judgements are not expressive of *beliefs*, properly understood). However, note that, once the background appeal to the philosophy of mind has done its work in filtering out some syntactic candidates for truth-assessment as spurious, a more-or-less disquotationalist account might still be given of the residual role of the truth-predicate *within* the remaining class of legitimate applications; and the content of adverbial modification of the truth-predicate will then still have to be explained by appeal to instances of the schema [M].

I claim that what goes for this kind of 'filtered disquotationalism' holds right across the class of relatively deflationist theories of truth – in particular, they all sustain the claims [A] and [Exp]. And, as we saw, other theorists will want to join the party. Indeed, prescinding now from questions about exactly which theories of truth positively *require* them, these two claims are surely attractive as stand-alone theses. And that is all that we really need here.

Third, to fend off a possible bad misunderstanding: though the framework approach to approximate truth encapsulated in [A] and [Exp] should particularly appeal to those already attracted by a deflationist account of plain truth, the approach *isn't*, and isn't intended to be, itself deflationist in any sense. All [Exp] tells us is what needs to be understood if we are to get a grip on the notion of approximate truth – namely how the sentence-adverb 'approximately' functions. But spelling *that* out in detail will require a substantial story, a rather different story in different cases. (Compare: there's a modal analogue to [A], namely

[P] 'P' is possibly true if and only if possibly P.

The deflationist about plain truth should again understand [P] as explanatory right-to-left. But making this point isn't being deflationary about possibility – all the substantial work of elucidating the modal operator evidently remains to be done.)

5.3 How are we to apply [A] and [Exp] to scientific theories? Suppose we present a theory as a conjunction of propositions: then the schema [A] blandly tells us that the theory 'A & B & C ...' is approximately true just if approximately (A & B & C ...). But now how do we move the approximation operator across the conjunctions? 'Approximately (P & Q & R ...)' is *not* generally equivalent to 'approximately P & approximately Q & approximately R ...', since the latter does not imply the former (suppose e.g. that P, Q and R report mutually incompatible estimates of some quantity, all close to correct). We will therefore need, in particular cases, some principled way of distributing the effect of the approximation operator across a theory-conjunction.

There is no general story to be told here: but for one class of theories, however, we can easily make some headway with the distribution problem. Let's say that a theory is a 'geometric modelling theory' – a *GM-theory*, for short – if it has the following two main components. Component *M* is to be purely mathematical, specifying a certain geometrical structure (e.g. by giving governing equations, or by specifying transformations for which the structure is the unique invariant). Component *A* is to give empirical application to the mathematically characterized structure by claiming that it replicates, subject to arbitrary scaling for units etc., a geometric structure to be read off some real-world phenomenon. Now, by [A], to say that a GM-theory comprising *M* and *A* is only approximately true is to say that, approximately, *M* & *A*. But presumably, in the normal case, we won't want to deny that the mathematical kernel of the theory is a perfectly correct characterization of some abstract geometric structure. In other words, the approximation operator is naturally taken to pass cleanly by *M* and apply only to the second empirical component of the theory-conjunction. In the general case, the component *A* may itself be conjunctive, and we will now be faced with a further distribution problem – but at least we've made a start.

So, the claim that a GM-theory is approximately true is naturally to be taken as the claim that the geometric structure in question approximately replicates the relevant structure to be found in the target real-world phenomenon.

And what, more carefully, is it for one geometric structure to approximate another? Different cases will require different treatment, depending on which features of the geometric structures are the prime focus of interest in a given context. The very simplest case is where the structures in question are both curves embedded in the same space: and the obvious thing to say (as a first shot) is that one such structure approximates to the other if the first curve can be distorted into the second by a transformation which (a) moves points by no more than some ϵ, and (b) preserves 'smoothness' (roughly speaking, first and perhaps higher derivatives at corresponding points are within some δ) – in such a case, one curve will closely 'track' the other. Bundles of curves can count as close if a suitable correspondence can be set up so that corresponding curves track within ϵ. And more complex structures will require more complex stories about 'closeness'. But while the geometric details may get tricky, as noted before there need be no *conceptual* problem about claiming that one structure approximates to another. Why should there be? – for it is precisely in these contexts where we can take metric ideas quite literally that the notions of closeness and approximation are most at home.

In short then, given schema [A], the claim that a certain GM-theory is approximately true should be conceptually entirely unmysterious: elucidating the claim will simply require spelling out what it is for an appropriate geometrical closeness relation to hold between structures of the relevant kinds. And the application of this abstract point to dynamical theories, chaotic and otherwise, is immediate. For such theories, as we have seen, are apt candidates for regimentation in GM form.

Take again the textbook account of the pendulum in GM form: this tells us that *M* (there is an abstract structure of a certain kind) and that in a good sense approximately *A* (approximately, the abstract structure replicates a structure encoding the real pendulum's spectrum of possible behaviours). Plausibly, when *M* is non-contingent, the inference from (*M* & *approximately A*) to *approximately* (*M* & *A*) is warranted; and so by an application of schema [A], it follows that the regimented standard account of the pendulum is approximately true. As it indeed is.

Precedents and comparisons

The emerging account of approximate truth for dynamical theories that invite the GM-theory treatment is still schematic: but the outlines are clear enough to enable comparisons with some general accounts of approximate truth as they would apply to the same cases.

But first, a brief historical aside. We are viewing dynamical theories from a stance which is rooted in work by Patrick Suppes (see e.g. Suppes 1960). He argued that the objects treated by (for example) textbook classical mechanics – the familiar cast of point masses subject only to gravitation, of massless springs, frictionless joints, and the rest – are really fictional ideals, best thought of abstract objects, which purport to mirror in their behaviour some class of real phenomena. So, on this account, various mathematically articulated theories are best regarded as having a kernel specifying a certain abstract structure (or set of structures), with the rest of the theory indicating application rules associating the structure(s) with certain worldly phenomena which supposedly exemplify them. What we have done, in effect, is simply to note that in the case of many dynamical theories, we can finesse any general problems about the talk of 'structure' here by giving a very literal-minded geometric reading of the notion. Indeed, we can *identify* e.g. an 'ideal pendulum' with a particular geometric structure in some appropriate abstract phase space: this is ontologically economical, as well as smoothing the way to an unproblematic account of approximate truth for dynamical theories.

The Suppes view of theories is now best known in a version which gives it a strongly empiricist twist: so it is worth stressing that empiricism is very much an optional extra. The logical empiricists treated theories as comprising a formal calculus and 'correspondence rules' relating some formal terms with observables. Suppose we take the formal calculus not as initially uninterpreted but rather (following Suppes) as specifying some formal structure(s); and take the correspondence rules to link not *terms* with observables but formal *structures* with observables. This yields an updated empiricist view of theories which is exactly Bas van Fraassen's. 'To present a theory,' he writes, 'is to specify a family of structures ... and to specify certain parts of these models as candidates for the direct representation of observable phenomena.' (van Fraassen 1980, 64) But it is one thing to hold that some mathematically framed theories can be regarded as having two components, specifying formal structures (e.g. geometric ones) and correlating these to the world: it is quite another thing to hold that the correlation between the textbook abstractions and the real world are to be confined to *observable* phenomena.

It is entirely consistent with our general approach to hold that theory-to-world correlations can link abstract geometric structures with (encodings of) the time-evolution of a range of physical quantities which includes *un*observables. There is nothing intrinsically empiricist, then, about our assumed GM treatment of dynamical theories.

How, then, does our story about approximate truth in the case of GM-theories relate to various general familiar stories about approximate truth for theories? The best account, due to David Lewis, is discussed in §5.4; here, I will very briefly comment on two other general approaches.

Suppose we have achieved a passable theory by ignoring the small effects of (as it might be) friction and air resistance. If we want a more accurate theory, we can try tinkering with our equations, adding the necessary correction terms. Complicating a theory in this way may well make it less mathematically tractable: that's why we will often prefer to stick with an inaccurate theory that idealizes by ignoring small perturbations. But in such a case – we might be tempted to say – our preferred inaccurate theory approximates to the truth in the sense that the theory *could* be developed into an accurate one by adding in enough small correction terms, hitting truth in the limit. (In other cases, no doubt, it is better to think of the approximately true theory not as something that can be made true by adding epicycles, but rather as something that can be derived by starting with the truth and then simplifying. Newtonian dynamics is approximately true, we might say, because it is what you get when you start from the relativistic truth and then, inter alia, set a small quantity – the reciprocal of the velocity of light – to zero.)

It may be suggested, then, that we deem a dynamical theory (taken as a set of equations) to be approximately true when it differs by small modifications, in one direction or the other, from a true theory. But this suggestion faces insuperable problems. For a start, as soon as we try to think in a principled way about what counts as a 'small' modification of equations, or try to make good the picture of there being a limit to the needed small adjustment terms, we discover that this story won't really do even in the cases where it looks most promising. But here I want to emphasize a different point – namely that there are central kinds of case which the story doesn't even begin to fit. Call a theory 'stubbornly unrevisable' if it *cannot* be turned into the truth by adding small correction terms (and cannot be derived from a true theory by small backwards modifications). Then a theory may still count as approximately true, by any sane standards that respect what we ordinarily say outside the philosophy seminar, even if it stubbornly resists revision. Theories developed within the framework of classical fluid mechanics provide obvious candidates. Such sets of equations cannot be made strictly true by fine-tuning, for the classical framework embodies the essential axiom that fluids are perfect continua – and no piling up of added epicycles is going to cancel that axiom and so deliver a theory which *is* strictly true of real, atomically granular, fluids. (Further, there is no backward path via easy simplification from e.g. a quantum statistical mechanics of molecular motion to a classical theory.) Hence classical fluid theories are stubbornly unrevisable; and yet some such theories – those that guide the design of aircraft wings, for instance – surely count as approximately true in their domain, if any theory does.

In sum, to assert that a theory is approximately true is to say that the *world* is roughly as the theory says it is, not to say that *another theory*, related by small adjustments, is precisely right. That is why a theory that stubbornly resists modification can still be approximately true: a theory might be true of a suitably nearby *world*, even if no 'nearby' *theory* is true.

So a 'small modifications' account of what makes for approximate truth won't do. Nor, I claim, will the much more developed line on approximate truth found in the work of Graham Oddie (1986) and Ilkka Niiniluoto (1987) – they are responding, in neo-Popperian vein, to the well-known failure of Karl Popper's original definitions of verisimilitude.

We need not pause to consider how far this latter approach is *intended* to cope e.g. with an application of the notion of approximate truth to dynamical theories; but I think that it is evident that in this domain at least, the neo-Popperian approach points us in entirely the wrong direction. Popper, recall, tried to define verisimilitude in terms of plain truth and plain falsehood (no approximation involved); roughly speaking, one theory has greater verisimilitude than another if it accurately hits the truth more often. Likewise for Oddie and Niiniluoto; when the wraps are off they are revealed as still wedded to the idea that in the end it is the number of bull's-eyes that matters. Their construction of a metric for distance from the truth begins from ideas like this: one basic conjunction of atomic propositions or their negations is nearer the truth than another if more of its conjuncts hit the truth. (The fun really starts when we try to work up to an account of distance from the truth for quantified languages: but the fundamental Oddie/Niiniluoto claim remains that 'the simple counting procedure that works so admirably in the case of finite propositional languages can be carried over to languages of any desired degree of complexity'. (Oddie 1986, 176)) So for them, getting near the truth is – at bottom – like getting a good score on a test asking a lot of yes/no questions: and getting *nearer* the truth is, at bottom, switching answers by inserting or deleting some negation signs.

This seems a hopeless initial model for approximate truth in the case of theories apt for regimentation in GM form. Most approximately true dynamical theories aren't theories that get a suitable proportion of basic claims dead right (other than the non-empirical, purely mathematical claims). Rather, they get enough things near enough right. And it seems particularly clear in the case of the false-because-over-intricate chaotic dynamical theories that they can't be repaired by anything remotely like the simple expedient of twiddling a few negation signs in the basics in some canonical formulation. Or that, at any rate, is the intuitive challenge to the Oddie/Niiniluoto approach. The approach might be worth pursuing further if it were the only game in town. But, as we've seen, it isn't.

5.4 A theory is approximately true if the way things actually are approximates to the way that the theory says they are – approximates, that is, to the way things would have been, had the theory been strictly true. Suppose we now make the familiar move from talk about 'the way things are' or 'the way things would have been' to talk about possible worlds. Then we might say: a theory is *approximately* true if the actual world is close to the possible world where the theory is *strictly* true (except that there will be many worlds at which a given theory is true: so we need to say, rather, that a theory is approximately true if the actual world is close to the set of possible worlds where the theory is true).

This sort of account has been endorsed by David Lewis (Lewis 1986, 24–27). Obviously the account needs fine tuning, by elucidating the correct closeness relation: is it simple 'separation at closest approach', or some more complex function of the distances between the actual world and the worlds where the relevant theory is true? But, relatively independently of the tuning, the account has evident virtues. For a start, the basic framework is familiar – possible worlds ordered by similarity. Why not buy the account?

Well, it certainly won't do in the simple, Mark I, version as just sketched. Take, say, a classical fluid theory approximately describing flow over an aerofoil. Any possible world of which such a theory is strictly true is one where fluids are perfect continua (that's the classical axiom). So, none of the fluid stuffs in our world is exemplified in that world and vice versa; and very many of the laws – for a start, those governing phase transitions from the fluid to the gaseous state – must be unimaginably different. Overall, the physics of such a world will be *radically* unlike that of the actual world. But this means that any world of which a classical fluid theory is strictly true will be nomologically a very remote world. So, assuming that we assign nomological considerations anything like their usual weighting in ranking similarities, the classical theory is not true at any *nearby* world – and hence (according to the Mark I possible world theory, however fine-tuned) isn't approximately true. That's a reductio, given that what we are after is precisely an explication of the pre-theoretic sense in which such theories *are* approximately true. (Likewise, given the infinite intricacy of their structures, chaotic theories can not be strictly true at any nearby world.)

Now, it might well be protested that this objection is rather a cheap shot and is very easily met. It is, after all, very familiar that different applications of the possible-world framework may require different standards of similarity. So in the present case we just need an appropriately qualified account of what makes for relevant similarity between worlds – an account which allows worlds to count as relevantly similar for the purposes at hand even though differing widely in other, perhaps very basic, nomological respects. But then how is the Mark II story to go?

There seems to be no prospect of an account that makes the kinds of similarity relevant to approximate truth independent of the theory whose approximate truth is up for consideration. Rather, the story has to run (very roughly): theory T is approximately true if the actual world and some world where T is true sufficiently resemble each other *in the key respects which T is concerned to describe*. Which respects to give weight to, and what to count as sufficiency in these respects, will no doubt be interest-relative. Thus classical fluid mechanics is approximately true because the actual world and some world where this theory is true resemble each other as far as the medium and large-scale behaviour of fluid flow is concerned (and nomological divergences in micro-respects don't matter). Classical genetics is approximately true because the actual world and some world where that theory is true resemble each other as far as the gross patterns in inherited characteristics are concerned. And so on.

However, while this revised possible world treatment of approximate truth may meet the earlier objection, it does so at the price of a certain circuitousness. For on anybody's view, full-blown possible worlds are richly structured. To talk of some world of which a given theory is true is in the general case to talk of something with vastly more detail than is required just for satisfying the theory. So if we move from a theory to a world of which it is strictly true, but then compare that world with the actual world only in the limited respects which the theory targets, we have in effect first added a great deal of unspecified structure and then carefully ignored it. Why go in for this round-about manoeuvre?

(Contrast possible world treatments of counterfactuals, where the invocation of full-blown worlds seems to be doing some real work. The fundamental insight is, roughly, that 'if P were the case, then Q' is true if the state of affairs P *along with potentially unlimited quantities of relevant background facts* suffices for Q. So giving truth conditions for the counterfactual in terms of what obtains at near P-worlds, with all their detail, is precisely a way of ensuring that all the required background gets into the story.)

Consider the special case of dynamical theories in GM form again. As a first shot, such a theory would be *strictly* true of some physical system if that system precisely exemplifies just the geometrical structure required by the theory. So the revised Lewisian account of approximate truth, applied to such a theory, comes to this: the theory is approximately true when the actual world closely resembles some other possible worlds wherein the target phenomena exactly exemplify the geometric structure required by the theory – where the resemblance is in the relevant respects which the theory addresses. But when will the actual world suitably resemble a possible world which exemplifies a given geometric structure? When the target phenomena in the actual world exemplify a structure which – in some appropriate sense – is close to the target structure. So the Mark II Lewisian theory entails that a theory is approximately true if the actual world exhibits some geometric structure sufficiently close in respects the theory cares about to the structure described by the theory. But note that this consequence is itself a *direct* account of approximate truth, the very one we have already given; the detour via other possible worlds has done no extra work.

Hence, applied to dynamical modelling, the crude Mark I Lewisian account of approximate truth is simply wrong; and the Mark II version, applied to dynamical modelling, entails our earlier account.

Miller's problem

There is a familiar argument due to David Miller (in e.g. Miller 1994, Ch. 11) which must now be faced: for, if sound, it would sink any account of approximate truth for GM-theories that invokes ideas like 'close-tracking'.

To take a simple illustration, suppose we have rival dynamical theories, H and K, which give values for the state variables x and y at time t. Then, Miller notes, we can have the following situation. (1) the trajectory predicted by K manifestly tracks the true values of x and y better than H. But (2) there is a transformation of coordinates, from x, y to x^*, y^*, and H now gives values for

x^* and y^* which manifestly track the true values of the new variables better than K's values. So, on the close-tracking account of approximate truth, we have: H is further from the truth than K in one coordinate frame, but nearer the truth in a second frame. But any plausible account of distance from truth should make comparative approximation to the truth coordinate invariant. And that's bad news for the close-tracking proposal.

Miller's own example is this: take

$$\begin{array}{lll} \text{H:} & x(t) = t & y(t) = 5t \\ \text{K:} & x(t) = t + a & y(t) = 2t \end{array}$$

and suppose that the true values of x and y are given by

$$x(t) = t + 2a \quad y(t) = t.$$

Uncontroversially, K tracks the correct values closer than H. Now take the coordinate transformation

$$\begin{aligned} x^* &= x + (5a/12t)y \\ y^* &= (5t/2a)x + y. \end{aligned}$$

The theories H and K become

$$\begin{array}{lll} \text{H:} & x^*(t) = t + 25a/12 & y^*(t) = 5t^2/2a + 5t \\ \text{K:} & x^*(t) = t + 22a/12 & y^*(t) = 5t^2/2a + 9t/2 \end{array}$$

while the true values of x^* and y^* are given by

$$x^*(t) = t + 29a/12 \qquad y^*(t) = 5t^2/2a + 6t.$$

And equally H, it seems, now tracks the correct values more closely than K. But H cannot both be nearer and further from the truth than K: hence close-tracking cannot constitute nearness to truth.

How seriously should we take this kind of example? The coordinate transformation required to reverse the fortunes of H and K is time-dependent in a pretty odd way. Consider: a constant unit line along the original x-axis becomes (applying any reasonable non-time-dependent metric in the new coordinate system) a line whose length grows with time; while a constant unit length along the original y-axis becomes a line whose length shrinks with time. With this kind of gerrymandered time-dilation, no wonder we can get strange results ('lengths' will vary without cause or effect, laws will cease to be time-invariant, and so forth). So the obvious initial response to Miller's actual example is to complain that the unnatural type of time-dependent coordinate transformation he invokes is illegitimate.

Still, while this complaint has intuitive force, it isn't immediately obvious how best to make the complaint stick. And in any case, banning weirdly time-dependent coordinate transformations won't resolve the matter. For consider a second example. Again there are two hypotheses

$$\text{H}': \quad x(t) = 5t \qquad y(t) = 3t$$
$$\text{K}': \quad x(t) = 4t \qquad y(t) = 2t$$

whereas the truth is that

$$x(t) = t \qquad y(t) = t.$$

Neither H′ nor K′ is a very good shot at tracking the actual values, but K′ is surely the better. However, take the coordinate transformation

$$x^* = x - 3y/2$$
$$y^* = y - 3x/4.$$

Then we have

$$\text{H}': \quad x^*(t) = t/2 \qquad y^*(t) = -3t/4$$
$$\text{K}': \quad x^*(t) = t \qquad y^*(t) = -t$$

and the true time-evolution is

$$x^*(t) = -t/2 \qquad y^*(t) = t/4.$$

And now H′ appears to track better than K′. So we have a coordinate transformation which in this case is nicely time-independent but which again reverses the fortunes of two hypotheses.

It might for a moment be wondered whether this phenomenon can only arise because we are considering cases where there are alternative hypotheses for *single* phase space trajectories, not hypotheses about whole *families* of trajectories from different initial states. So is there safety in numbers or can there still be Miller-type reversal between generalized hypotheses?

There can. Just generalize the last example. Suppose our rival hypotheses say that, for any initial values $x(0) = a$, $y(0) = b$ the resultant trajectories will be

$$\text{H}'': \quad x(t) = 5t + a \qquad y(t) = 3t + b$$
$$\text{K}'': \quad x(t) = 4t + a \qquad y(t) = 2t + b$$

whereas the truth is that

$$x(t) = t + a \qquad y(t) = t + b.$$

K″ is better than H″ on the time-evolution of the values of x and y from any initial state: but, with the same transformation, it is immediate that H″ is better than K″ on the time evolution of x^* and y^* from any initial state.

Recall, however, a point already stressed – that what makes for approximate truth is interest-relative: a theory counts as approximately true if it close-tracks the world in some respects we care about. Suppose for a moment that in our last example x and y respectively represent, as it might be, temperature and circulation velocity in a steadily heated convected liquid, and so H″ and K″ are hypotheses about how these quantities increase over time. Then x^* represents temperature minus 150% of the circulation velocity. And what – we may well ask – is the real physical significance of *that*?

Objection: the question is a cheat, for 'temperature minus 150% of circulation velocity' isn't a dimensionally coherent quantity. Reply: fair point, but just suppose the H/K pairs are doctored with suitably dimensioned multiplicative constants of unit value so that dimensional coherence is restored. Then the point remains: there is, in the general case, no reason to suppose that, if x and y are physically significant quantities, x^* and y^* will be too. But if x^* and y^* aren't of interest, why care whether a theory close-tracks them? Dynamical theories aim to track the time evolution of physically significant quantities, and a theory will count as approximately true just so long as it gets the values of *those* quantities near enough right for long enough (so e.g. K″ counts as approximating the truth better than H″ since it is more accurate on the values of the physically significant quantities).

Does talk of certain properties as 'physically significant' require giving a special metaphysical status to certain 'elite' properties (in the manner of David Lewis)? Not obviously so. It seems enough for our purposes to note that, in real cases, the quantities represented by variables in a particular dynamical theory – temperature and circulation-velocity, in the case we just imagined – will (according to other theories) feature in a wide range of functional relationships to other quantities such as pressure, volume, viscosity, etc. And while the $\{x, y\} \rightarrow \{x^*, y^*\}$ transformation is tailor-made in order to transform the H and K pairs of hypotheses so as to preserve functional simplicity while reversing intuitive accuracy, there is no reason at all to suppose that the same transformation (or any extension thereof to include variables over other quantities) will produce anything but mess when it operates over a wider domain of hypotheses. So in the setting of a wider physics, standard simplicity considerations can privilege a certain choice of quantities as the 'physically significant' ones.

So the state of play is this. In the general case, to get the Miller reversal phenomenon, we will have to use coordinate transformations between phase spaces which are time-dependent in a quite anomalous way. Only in special cases can well-behaved transformations lead to reversal. And in most of those special cases, we can still reasonably prefer working in one phase space rather than another on the grounds that the state variables in the one case but not the other represent physically significant quantities (as evidenced by other theories). So the only case we would really have to worry about is the case where transformations which were not time-dependent in an anomalous way took us from one space to another, reversing the fortunes of some hypotheses, but where the state variables of both spaces had equal claim to physical interest.

I know of no such case: but if there were one, we could just live with it – the relevant hypotheses H and K would, as it were, score one goal each, so the match would be a score draw, and neither preferable. An account of approximate truth doesn't have to rule out such cases.

5.5 We have seen, then, that [Exp] combined with a (re)conceptualization of dynamical theories as GM-theories yields a very natural account of their approximate truth in terms of suitable geometrical approximations. And there is an attractively simple proposal about what might constitute geometrical closeness in some basic cases – namely, accurate tracking of phase-space trajectories.

To emphasize: it is *not* being claimed that, across the board, truth in a theory is just accuracy. The proposal is a limited one, geared to GM-theories of modest ambition, which just aim to tell us how certain quantities evolve over time. It is only for these that close tracking suffices for approximate truth.

What of the following objection? Suppose we have a pair of dynamical models of planetary motion, where T_1 is Ptolemaic and T_2 Newtonian. It could well be that the parameters in T_1 and T_2 are so chosen that in fact T_1 tracks the actual behaviour of the planets better than T_2. But must we say that the Ptolemaic T_1 will therefore be nearer the truth than T_2? We have allowed that judgements of approximate truth will be interest-relative: and if our concern is (say) merely navigational – if we just want to predict the position of the planets against the fixed stars – then T_1 will by hypothesis indeed be nearer the truth in the respects we then care about. Surely there are perspectives from which the Newtonian T_2 (still conceived as a GM-theory) should rate as nearer the truth. The objection is that a 'geometric closeness' account cannot allow us to say this.

But not so. For the Newtonian T_2 scores when it is seen as an instance produced according to a more general recipe which unifies many other dynamical theories, where enough of these other theories *are* approximately true in a basic tracking sense. Even if we have wrongly set the parameters so that T_2 doesn't track the world very well, it can still get the reflected credit (as compared with a Ptolemaic model) of belonging to a unified family with many successful trackers. Let's grant, then, that we may count a Newtonian dynamical model as approximating the truth even though it doesn't itself track the relevant quantities so well: but at bottom it can still be close tracking (albeit of other quantities by other specific dynamical theories in the family) that underpins the judgement which rates T_2 nearer the truth than T_1.

Thus, with modest embellishment, the closeness-as-tracking proposal can cope with the sketched objection. Other cases will require further work. In particular – to return at long last to our main concern – we saw that, in the cases of applied chaotic theories, we may care less about the close tracking of individual trajectories (for even when there is successful close tracking, exponential error inflation means that there are low limits to our ability to exploit this accuracy in forming useful trajectory predictions). We are likely to emphasize instead certain more abstract geometric similarities between the dynamical models and structures in worldly behaviour. We may care, that is to say, about similarities in such things as bifurcation behaviour, the structure of 'routes to chaos', the behaviour of Liapunov exponents (which measure the rates at which trajectories spread apart), and the like. And we will count the chaotic theories as getting near the truth in respect of *these* prioritized similarities. Further, we may now want to stress more emphatically the 'smudging' of the relevant geometric structures found in the world. More of the story will emerge in the next chapters. But for now the important point is this: the relevant similarity relations between chaotic structures and worldly ones which underpin talk of approximate truth here are still geometric. So, however the technicalities go, the application of the notion of approximate truth to an applied chaotic theory is still conceptually unproblematic.

6

Universality

6.1 An applied chaotic dynamical theory claims, in essence, 'the time-evolution of the relevant real-world quantities has *this* intricate structure'. As we have seen, despite its non-empirical surplus content, such a theory can count as approximately true; and, despite sensitive dependence on initial conditions, it can be richly predictive. It will not be straightforwardly causal-explanatory (§5.1): but that does not stop the theory being explanatory in other ways. And our next major task is to say something about the sorts of explanations that chaotic dynamics can deliver.

As background, we need to take on board rather more of the elementary mathematics of chaos. That's the business for this largely expository chapter. Careful exposition will, in particular, remove the appearance of number-magic that can attach to chaotic dynamics (§6 below). As usual, the headline news is in the main text; the purely mathematical interludes here fill in a few details and sketch some proofs, but can be skipped.

We begin by turning our attention away from the Lorenz system to another standard exemplar of chaos, the so-called 'logistic map', which is a discrete map defined over the unit interval. At first sight, this might look like a radical change of tack. So far, we have been discussing the behaviour of phase space trajectories that are solutions to sets of linked first-order differential equations in the standard form (D). But the logistic map is a simple *difference equation*,

$$x_{n+1} = \mu x_n(1 - x_n), \quad x_i \in [0, 1], 0 \leqslant \mu \leqslant 4$$

which gives x_{n+1}, the value of the variable x at the $n + 1$th check-point (maybe, at $n + 1$th tick of the clock), in terms of the previous value x_n. So this map defines not a continuous trajectory but a discrete *orbit*, i.e. a sequence of values $x_0, x_1, x_2, \ldots$.

There is an intimate connection, however, between discrete orbits and continuous trajectories. Consider again our simplified model of the dynamics in the Lorenz system (§4.2). In that stripped-down version, trajectories keep returning to cross a 'base-line'; and so the continuous dynamics of a trajectory crossing this line at x_0 produces a sequence of successive base-line hits at x_1, x_2, x_3, etc., yielding the map $x_0 \Rightarrow x_1 \Rightarrow x_2 \Rightarrow x_3 \ldots$. And indeed, we found it easiest to understand the behaviour of a trajectory in the model by focusing on the orbit formed by these successive crossing-points.

This tactic of illuminating the behaviour of a continuous model by considering a related (one-dimensional) discrete map works much more generally. Take first a continuous dynamical model in three variables, and imagine cutting across the 3D trajectory bundle with a suitable 2D surface *P*. Then a trajectory passing through the point x_0 on the surface *P* winds round (maybe at irregular time intervals) to hit this surface again, first at x_1, then at x_2, and so on, inducing a 2D discrete recurrence map $x_n \Rightarrow x_{n+1}$ among crossing-points on *P*.

But now recall that, in the chaotic systems that are our concern, we are mainly interested in what happens in the close vicinity of the

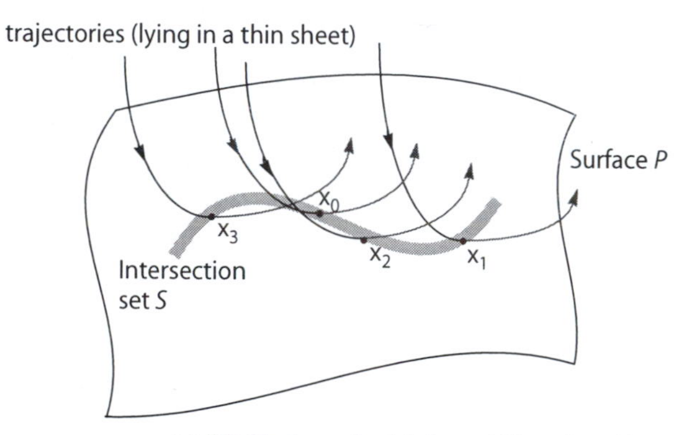

(a) 3D Trajectories hitting a 2D surface *P*

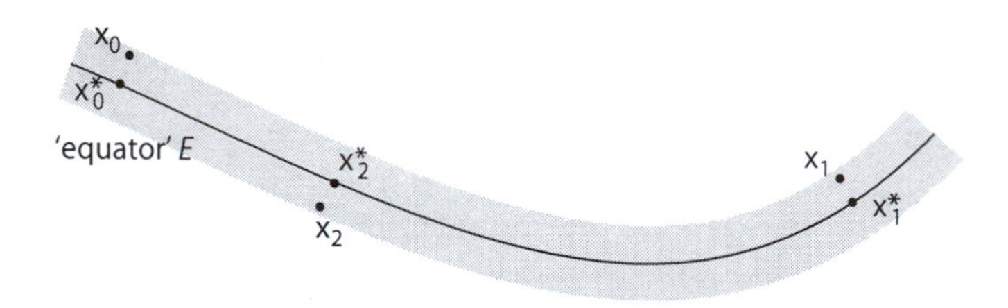

(b) The 2D map $x_0 \Rightarrow x_1 \Rightarrow \ldots$ inducing a similar 1D map $x^*_0 \Rightarrow x^*_1 \Rightarrow \ldots$ on the 'equatorial' curve *E*

Figure 6.1 From 3D trajectories to a 1D map

attractor. Trajectories hereabouts must overall get scrunched together ever closer to the attractor: but since there is sensitive dependence on initial conditions, trajectories must also diverge apart. How can we have both compression and spreading? In some typical cases, like Lorenz's, by the trajectories locally spreading apart in one dimension while being squeezed together in other dimensions to form 'sheets', which are then folded back on themselves. Hence we can often choose a surface P to cut transversely across the 'sheets' – so that P intersects trajectories on or near the attractor in such a way that the crossing points form a set S which is stretched out thinly along the direction of trajectory-spreading (S falls in a region roughly like a smudged out curve as in Figure 6.1a).

And now consider the behaviour of the recurrence map $x_n \Rightarrow x_{n+1}$ for points in this set S. If we are concerned with the gross behaviour of this map, then little will be lost if we pretend that S actually has zero 'thickness' (if we identify the set S with, so to speak, an 'equatorial' curve E). In other words, the interesting behaviour of the recurrence map on S will be captured if we just ignore the wobbles above and below the equator, and pretend that this is a 1D map $x^*_n \Rightarrow x^*_{n+1}$ defined for equatorial points (see Figure 6.1b).

The salient features of the behaviour of 3D trajectories near the attractor may thus be reflected in the behaviour of a 1D map. And these considerations will generalize further. Even in an n-variable chaotic case, phase space trajectories are often squeezed together in all-but-one dimensions. And the resulting attractor will again intersect a suitably chosen $(n-1)$-dimensional 'Poincaré surface of section' P in a set like S which is almost one-dimensional (i.e. contained in a region like a curve slightly smudged out in the other $n-2$ directions). So the recurrence map formed by successive hits as a trajectory passes through S will again behave almost one-dimensionally.

In sum, even though our prime concern is with continuous systems defined by n-variable differential equations, we have good reason to suppose that the study of one-dimensional maps can be illuminating.

But why choose the logistic map for special attention? It turns out, and indeed this is a central theme of the chapter, that some of the key features of the dynamics of this map are 'universal' in the sense of being shared by a wide class of cases: and so, given its simplicity, the logistic map makes an nice exemplar. However, there is more to the choice than that.

Consider how a sheet of trajectories can both spread apart to yield sensitive dependence, and fold back to more-or-less rejoin itself to yield confinement. A simple way to achieve this is by 'horseshoe folding' – see Figures 3.2c, 3.2d (if these diagrams are not self-explanatory, then this is the time to tackle the associated interlude for a little more explanation): the effect is illustrated again in Figure 6.2. Focus now on the x-coordinate addresses of the successive points where trajectories pass through the cross-section, and pretend that (as in our model of the Lorenz model, §4.2) the trajectories spread apart in a uniform way and also that the thickness of the bend is negligibly small. Then a trajectory through a point x_n where $0 \leqslant x_n \leqslant 0.5$ will return to the point x_{n+1} with coordinate $2kx_n$, and a trajectory through x_n where $0.5 \leqslant x_n \leqslant 1$ will next visit the cross-section at $x_{n+1} = 2k(1 - x_n)$. The resulting map $x_n \Rightarrow x_{n+1}$ is called, for obvious reasons, a 'tent' map. And from here, it is a fairly small step on to the logistic map. For suppose now we imagine – slightly more realistically – that the stretching and folding produces a map whose graph lacks a sharp vertex, so we get a smoothed-out version of the same sort of 'one hump' recurrence map. So the tent will be replaced by something like a parabola peaking with the value k when $x = 0.5$. The logistic map for $x \in [0, 1]$ (with μ set at $4k$) is just such a parabola.

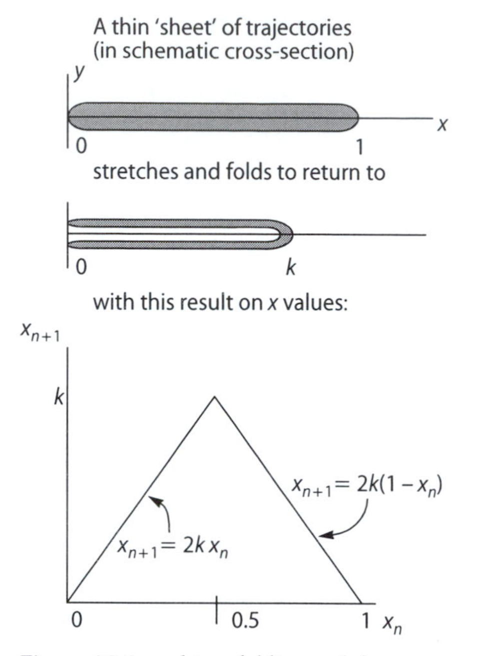

Figure 6.2 Stretching, folding and the tent map

So much, then, by way of plausible motivation for looking at the logistic map. Now to explore some of its complexities.

6.2 We will say that an orbit is *periodic* with period m if $x_{n+m} = x_n$ and there is no $l < m$ such that $x_{n+l} = x_n$. In other words, a period m orbit cycles round repeating itself after m steps, but no sooner. (An orbit is *eventually* periodic if there is some r and m such that $x_{n+m} = x_n$

once $n > r$.) Points on a period m orbit are said to be *period m points*. If x_0 initiates a period 1 orbit, so that $x_0 = x_1 = x_2 = \dots$, then we will say that x_0 is a *fixed point* of the map.

The fixed points of the logistic map are the solutions to $x = \mu x(1 - x)$, i.e. $x = 0$ and $x = p = (\mu - 1)/\mu$. When $\mu \leqslant 1$, $x = 0$ is the only fixed point within [0, 1]. And for any orbit starting in the interval at $x_0 \neq 0$, the x_n get asymptotically ever closer to 0 – i.e. every orbit is attracted to the fixed point. When $1 < \mu < 3$, the behaviour of orbits is again quite simple. The orbit starting at $x_0 = 0$ must remain at this fixed point. But the origin is no longer an attractor: orbits starting anywhere else in the unit interval now gravitate towards the other fixed point at p.

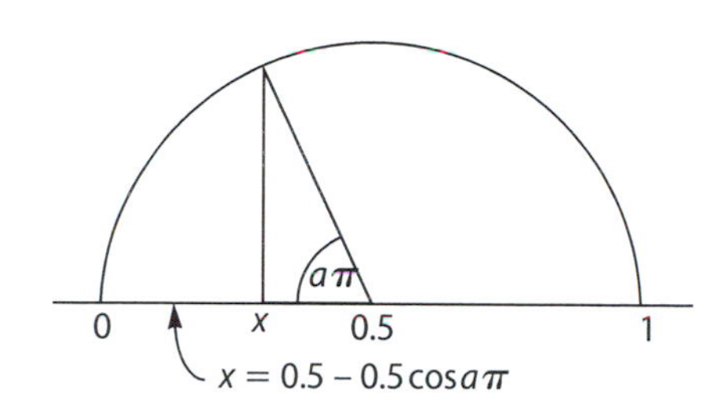

Figure 6.3 The unit interval parametrized by *a*

So far, so unexciting (and we noted in §1.3 that in *continuous* one-dimensional models nothing more interesting can occur). However, we can get radically more complex behaviours in the discrete case. Consider what happens when $\mu = 4$, so that we have

$$x_{n+1} = 4x_n(1 - x_n).$$

A neat trick simplifies the analysis of the behaviour of orbits here. Note again that x_n is always in the unit interval if x_0 is; and inspection of Figure 6.3 shows that there will be a unique a_n in [0, 1] such that

$$x_n = 1/2.(1 - \cos a_n\pi).$$

In terms of the a_n, the map thus becomes

$$\begin{aligned}\cos a_{n+1}\pi &= 2\cos^2 a_n\pi - 1\\ &= \cos(2a_n\pi).\end{aligned}$$

Solving (and remembering that the a_n are no greater than 1) we get

$$\begin{aligned}a_{n+1} &= 2a_n \text{ for } a_n < 0.5,\\ a_{n+1} &= 2(1 - a_n) \text{ otherwise.}\end{aligned}$$

This is a tent map on [0, 1]; and its behaviour is readily understood. For consider the binary expansion of a_n, $.s_1s_2s_3s_4s_5\dots$. If the initial digit s_1 is 0, then $a_n < 0.5$, and so $a_{n+1} = .s_2s_3s_4s_5s_6\dots$; there is a simple symbol-shift. While if s_1 is 1, then $a_{n+1} = .\hat{s}_2\hat{s}_3\hat{s}_4\hat{s}_5\hat{s}_6\dots$ (where $\hat{s} = 0$ if $s = 1$,

$\hat{s} = 1$ if $s = 0$); i.e. we have digit-flipping plus a shift. This is a minor variant on the simple symbol shift dynamics we looked at before, in §4.2. So similar reasoning shows that the orbits of the a_n are sensitively dependent on initial conditions. Rather than converging on a fixed point, nearby orbits diverge exponentially: if a_0 and a^*_0 differ by ϵ, then a_n and a^*_n will differ by $2^n\epsilon$ (at least for small enough n – once a_n and a^*_n straddle $a = 0.5$, orbits can get thrown together again). And though there is a denumerable infinity of (eventually) periodic orbits densely scattered through [0, 1], that still leaves continuum-many never-repeating orbits. Given that there is a straightforward, continuous, $a_n \Leftrightarrow x_n$ link, what goes for the tent map must go for the original logistic map with $\mu = 4$: so here too, there must be a chaotic mix of sensitive dependence, confinement, and non-periodic orbits.

6.3 In summary: for $0 \leqslant \mu < 3$, the logistic map has a very simple dynamics – orbits being attracted to one or other of the fixed points $x = 0$ or $x = p$. And for $\mu = 4$ nearly all orbits wander never-endingly all over the unit interval. Which raises the question: how does the chaos develop out of the initially simple dynamics as μ increases?

We can get a sense of the startling complexity of the story here by exploring using a computer. Pick a value of μ: then for arbitrary x_0 (say $x_0 = 0.5$) calculate the sequence of iterates $x_0, x_1, x_2, \ldots$. Throw away the first hundred or so members of the orbit – in other words, ignore the effect of any initial transients and look at the orbit once it has settled down towards its long term pattern of behaviour. Then plot the next few hundred iterates against the value of μ. If the orbit is attracted to a fixed point, then (having ignored any initial wandering around, and given the limited resolution of the computer calculation), we'll be plotting more or less that one fixed point. If the orbit gravitates towards a certain period m orbit, we'll see those m points plotted against the value of μ. If the orbit wanders chaotically over some interval, then a whole range of points will be plotted.

Having plotted the orbit of x_0 for one fixed value of μ, now repeat the exercise, incrementing the value of μ by steps of (say) 0.001 within [3, 4]. The result is Figure 6.4 – another familiar icon of chaos theory.

What do we learn from this orbit diagram? At $\mu = 3$, the fixed point p, which for lower values of μ attracted orbits, becomes repelling – nearby orbits diverge, so the fixed point no longer shows up on the plot of 'typical' orbit behaviour. Nearly all orbits now gravitate instead towards a period two orbit oscillating between a pair of points q_1 and q_2 (where these period 2 points separate as μ increases). At $\mu \approx 3.450$,

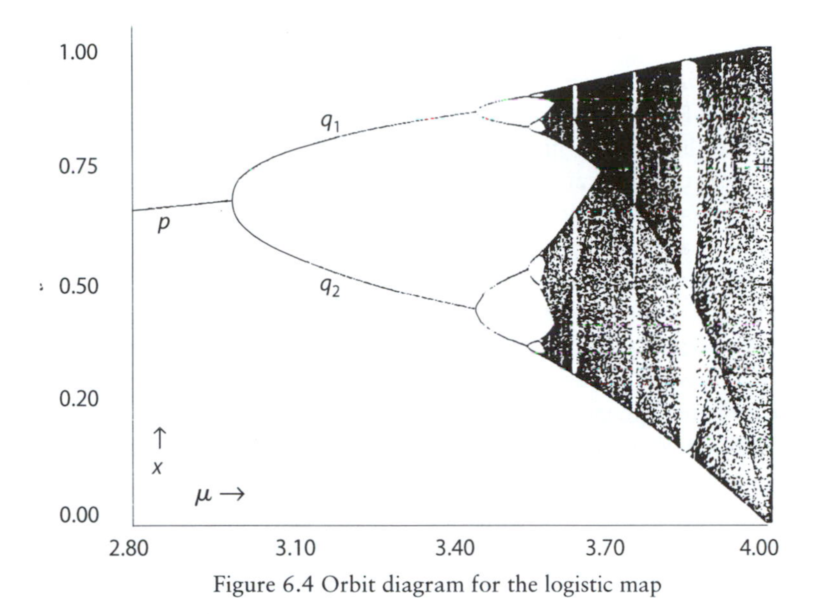

Figure 6.4 Orbit diagram for the logistic map

there is another 'bifurcation', the q_1/q_2 orbit ceases to attract, and a new attracting period 4 orbit is born. Then at $\mu \approx 3.544$ this period 4 orbit in turn becomes unstable, and the new attractor is a period 8 orbit. Further period-doubling bifurcations arrive ever faster (as a more fine-grained computer investigation reveals), and by $\mu = c_\infty \approx 3.570$ the attractor becomes an infinite set – in fact, analytic investigation shows that it is a Cantor set.

As μ further increases, there are more complications. For example, there is an interval of values for μ where there is an attracting period 3 orbit (beginning at $\mu \approx 3.829$, this window in the chaos is clearly visible in the Figure). Exploring the upper end of this period 3 window with finer increments for μ reveals that this orbit breaks up in another cascade of period-doubling bifurcations, leading to orbits of period 6, 12, 24 etc. Before the period 3 window, a period 5 window is also just discernible; and more detailed explorations reveal narrower windows with higher odd-number periods, which again break up by period-doubling.

And this is only the very beginning: there is a lot more structure yet to be found in the behaviour of possible orbits for different values of μ. For our purposes, though, we can concentrate mostly on that first period-doubling cascade.

'Period three implies chaos'

There is an issue about what exactly is revealed by the sort of computer trial just described. For remember that computer calculations have limited resolution and introduce round-off errors. Suppose that for a given value of μ there is a periodic orbit which attracts its neighbours (it can be proved, though we won't do so here, that when such an orbit exists for the logistic map, it will be unique). Then any co-existing *non*-attracting orbits will be invisible to the computer. Even if our chosen x_0 happens to initiate one of those unstable orbits – i.e. an orbit such that nearby ones diverge away from it – the chances are that the *calculated* orbit will quickly depart from the unstable *real* orbit as the round-off errors accumulate. And then the calculated orbit will wander around until pulled in by the attracting orbit.

When we are in a regime where *all* orbits are unstable – as happens e.g. when $\mu = 4$ and neighbouring orbits tend to peel apart exponentially – then all calculated orbits will be 'untrue'. But there are shadowing theorems which guarantee that a noisy computer calculation will closely track *some* orbit, even if not the one actually initiated at x_0 (see §4.4). So the calculation again hopefully at least reveals something about a 'typical' orbit – though strict proofs that shadowing orbits behave typically are in fact hard to come by.

Computer trials need, then, to be interpreted with some caution. In particular, to repeat, they won't reveal anything about the non-attracting orbits 'hidden' by attracting orbits. However, there is a quite astonishing theorem by Sarkovskii that tells us something about what invisible orbits there must be.

Order the integers as follows: list all the odd numbers from 3 upwards; then append 2 times these odd numbers, then 2^2 times the odds, then 2^3 times the odds, and so on. That leaves over just the powers of 2, which we tack on in descending order. This gives the Sarkovskii ordering:

$$3 \prec 5 \prec 7 \prec \ldots \prec 3{\cdot}2 \prec 5{\cdot}2 \prec 7{\cdot}2 \prec \ldots \prec 3{\cdot}2^2 \prec 5{\cdot}2^2 \prec 7{\cdot}2^2 \prec \ldots$$
$$3{\cdot}2^3 \prec 5{\cdot}2^3 \prec 7{\cdot}2^3 \prec \ldots \prec 2^3 \prec 2^2 \prec 2 \prec 1.$$

Let f be a continuous function from R to R. Then Sarkovskii's theorem states: if f has an orbit of period m, then it also has an orbit of period k for every k such that $m \prec k$.

Note that *all* that is assumed about f is that it is a continuous real-valued function defined over the reals. This very modest assumption holds, in particular, in the case of the defining function for the logistic map for any μ, so Sarkovskii's theorem applies. Hence, for example, for values of μ where our computer trial shows up an attracting period 3 orbit, there must also be orbits of *all* possible periods (and indeed an infinite number of infinite, aperiodic, orbits too), orbits which are computationally invisible because not attracting.

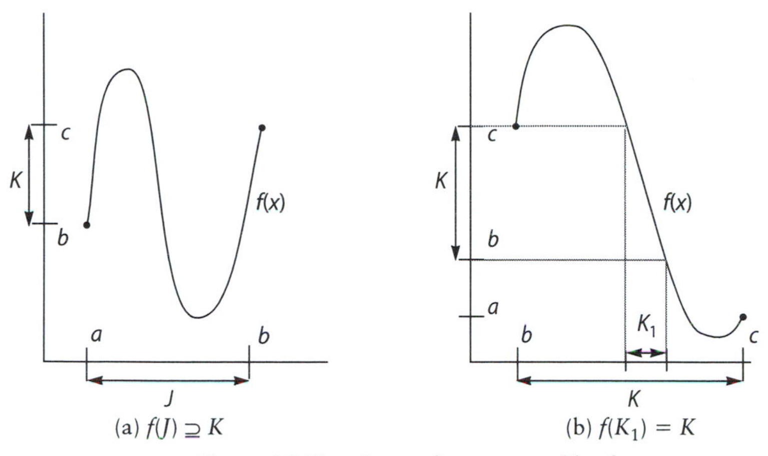

Figure 6.5 How intervals are mapped by f

This particular consequence that a period 3 orbit entails the existence of an orbit of every period was independently proved by Li and Yorke in their 1975 paper 'Period three implies chaos', whose title seems to have been as responsible as anything for introducing the current use of the term 'chaos'.

In the remainder of this interlude, we give a very brisk proof sketch of the weaker Li-Yorke result. This is mostly for fun, though a general moral is drawn in §6.7. We are concerned then with a function f which is continuous and which has a period-three orbit – i.e. there is a cycle of points a, b, c such that $f(a) = b, f(b) = c, f(c) = a$. And without loss of generality, we can assume $a < b < c$. We need to show that f has cycles of every possible period. For brevity, we will set $J = [a, b], K = [b, c]$; and if I is an interval, then we write $f(I)$ for the set of points $f(x)$ for $x \in I$. We will write $f(f(f(\ldots f(x))))$, i.e. the result of re-applying the function n times, as $f^n(x)$.

The continuity of f implies that $f(J) \supseteq K$. For by hypothesis f takes the end-points of $J = [a, b]$ to the end-points of $K = [b, c]$; and the intervening points in K are visited at least once as f varies between a and b (that's the Intermediate Value Theorem: see Figure 6.5a). Similarly $f(K) \supseteq [a, c] = J \cup K$. Now, $f(K)$ 'covers' K, so – by continuity again – there must be at least one closed sub-interval K_1 within K such that $f(K_1) = K$ (see Figure 6.5b). And since $f(K_1)$ covers K_1, there must by the same reasoning be a sub-interval K_2 within K_1 such that $f(K_2) = K_1$. Repeat n times, and we can construct a series of nested intervals $K_n \subset K_{n-1} \subset \ldots \subset K_1 \subset K$ such that $f(K_i) = K_{i-1}$. This means that n applications of f inflate K_n to K – i.e. $f^n(K_n) = K$. And since $f(K) \supseteq J \cup K$, $f^{n+1}(K_n)$ covers J. So there must be some $K_{n+1} \subset K_n$ which after $n + 1$ applications of f gets inflated exactly to J. And since $f(J) \supseteq K$, it follows in turn that $f^{n+2}(K_{n+1}) \supseteq K \supseteq K_{n+1}$. But

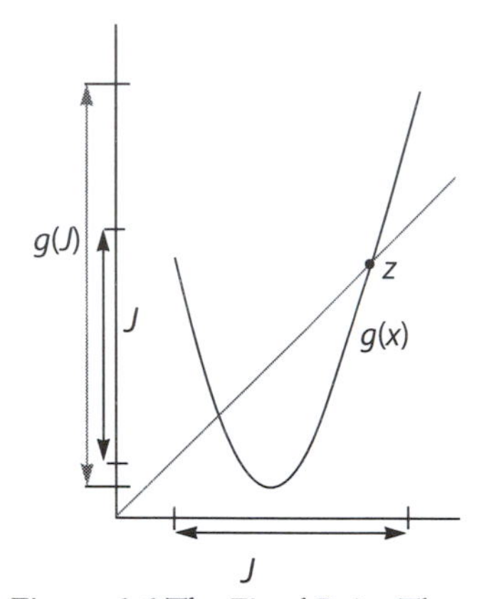

Figure 6.6 The Fixed Point Theorem

it is a familiar result that if $g(J) \supseteq J$, then g has a fixed point z in J where $g(z) = z$ (that's the Fixed Point Theorem, see Figure 6.6). So f^{n+2} has a fixed point p in K_{n+1}.

What happens to p as f is applied? By the construction of K_{n+1}, p first gets mapped to a point in K_n, then to one in K_{n-1}, then eventually to a point in K_1 then to a point in K, then to a point in J, then back at last to p itself, after $n + 2$ iterations (since p is a fixed point of f^{n+2}). None of the earlier iterations can take p back to itself (or else the orbit would never reach J). So p has period *exactly* $n + 2$. Hence we have shown that for any $n \geq 0$ there is a periodic orbit of period $n + 2$. By the fixed point theorem again, it is immediate that f has a fixed point orbit of period 1. So f has orbits of every period. QED.

It can also be established that – on the same assumption that there is a period three orbit – there must in addition be an uncountable number of non-periodic orbits in $[a, c]$ alongside all the orbits of finite period. Further, it can be shown (though we will not do so here) that at least in the $[a, c]$ region there will be sensitive dependence on initial conditions with iterates of nearby points spreading apart. So period three does indeed imply chaos. Proving the full Sarkovskii theorem is rather harder, though it still only requires relatively elementary materials.

One final point. Note that – applied to a function-family involving a parameter μ, such as the logistic map – Sarkovskii's theorem shows something about the co-existence of other orbits when a period n orbit shows up for a given μ. It doesn't in itself show anything about what happens for *other* values of μ, and in particular it doesn't say in what order the stable orbits appear as μ varies (though with more work, some such connections can be made in 'nice' cases).

6.4 In §1, we showed how the behaviour of one-dimensional maps may illuminate continuous dynamics; in §§2 and 3, we noted some of the behaviour of one map in particular. In this section, we start generalizing, by noting that certain key features of the logistic map are in fact *universal* features, shared across a large family of cases.

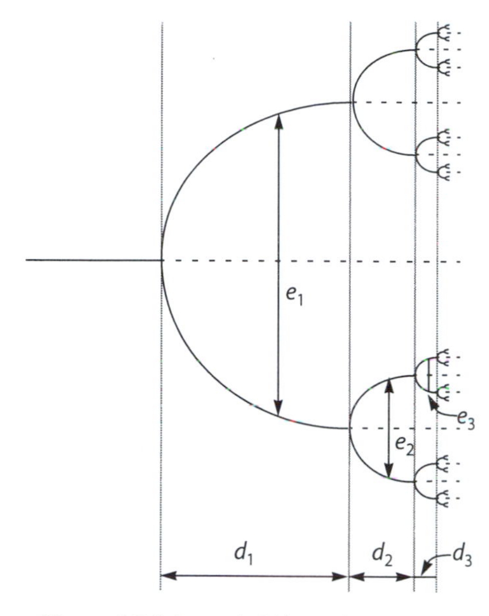

Figure 6.7 Schematic bifurcation structure (dotted lines: unstable periodic orbits)

Consider again the onset of period-doubling in the logistic map after $\mu = 3$. The structure revealed in Figure 6.4 is schematically represented in Figure 6.7 (with some vertical distortion to bring out the self-similarity in the diagram).

Let the interval between the values of μ where the n-th and $(n+1)$-th bifurcations occur be d_n. Then the intervals d_i approximate a geometrically decreasing series: the ratio d_i/d_{i+1} – call this the bifurcation rate – approaches a limit value $\delta = 4.6692...$ as i increases. δ is the 'Feigenbaum constant', so-called in honour of its discoverer. The ratios between the vertical spreads of the successive pitchforks in the orbit diagram also form a geometrically decreasing series. This relationship is rather more difficult to state precisely, because the pitchforks for a particular value of μ vary in height – see Figure 6.4 again. But there is a principled way of picking a standard indicator e_n of the spread of the pitchforks after the n-th bifurcation, and then the ratio of e_i/e_{i+1} tends in the limit to a second Feigenbaum constant, $\alpha = 2.5029...$.

Since the series of values of the parameter μ where there is a period-doubling bifurcation pile up ever faster, the series has a limit point $c_\infty \approx 3.570$. What, then, is the dynamics when $\mu = c_\infty$? In other words, what happens at the *end* of a period-doubling cascade?

Consider again the regime where there is an attracting 2^n orbit. That attractor doesn't attract *every* orbit: remember, the period $2^0, 2^1, 2^2, \ldots 2^{n-1}$ orbits still remain in existence as unstable, ghostly presences, no longer attracting their neighbours. And, as Figure 6.7 indicates, between any two points on the 2^n point attractor, there are points belonging to unstable orbits of lower period. Something of this behaviour persists when $\mu = c_\infty$: there is now an aperiodic attractor, but between any two points on the attractor there are points belonging to some (now unstable) period 2^n orbit. Hence the attractor can contain no intervals, and in fact it is another Cantor set: it turns out to have fractal dimension about 0.54.

Hence, at $\mu = c_\infty$ we have confinement (trivially) plus a complex mix of periodic and aperiodic orbits. What of sensitive dependence on initial conditions? Well, it can be shown we don't get this at $\mu = c_\infty$. But as soon as μ is even slightly greater than c_∞, orbits starting close together do peel apart exponentially. So when μ is just greater than the critical value, we do have the full signature of chaos, including sensitive dependence. Thus period-doubling is indeed a 'route to chaos'.

So much for the initial period-doubling behaviour in the logistic map. But – and here's the crucial observation – the very same basic pattern of period-doubling bifurcations piling up and leading to chaos turns out not to be unique to the logistic case. In fact, take *any* family f_μ of 'unimodal' maps of an interval into itself (where μ is a parameter which features in a suitably simple way, and a function is unimodal if – like the logistic map's defining function – its graph has just one hump in the interval, i.e. if the first derivative of the map has a single zero). Then we will again find a period-doubling cascade as μ varies. And perhaps much more surprisingly, these bifurcations will arrive at the very same limiting bifurcation rate δ and the pitchforks will spread out at the same limiting rate α. The Feigenbaum constants are indeed constants – values that re-appear in the description of period-doubling in a wide range of cases.

There are many other universal features of unimodal maps: but we will concentrate on the features of the initial period-doubling, for two reasons. First, it is not hard to understand why period-doubling should be so prevalent – and the next interlude gives some quick-and-dirty qualitative explanations for those interested (though it is a *lot* harder to prove the quantitative results, e.g. to show why δ should be a universal constant: for our purposes, quantitative universality can perhaps just be taken as a background fact from pure mathematics). And second, period-doubling is dramatically displayed in various empirical phenomena. What needs to be explained therefore, and so our next main topic, is how the universal features of a family of discrete maps are related to the modelling of real-world continuous processes.

More on period-doubling

Take the logistic map again: we can give an illuminating graphical analysis of the behaviour of orbits by means of a 'web' diagram as in Figure 6.8. Such a

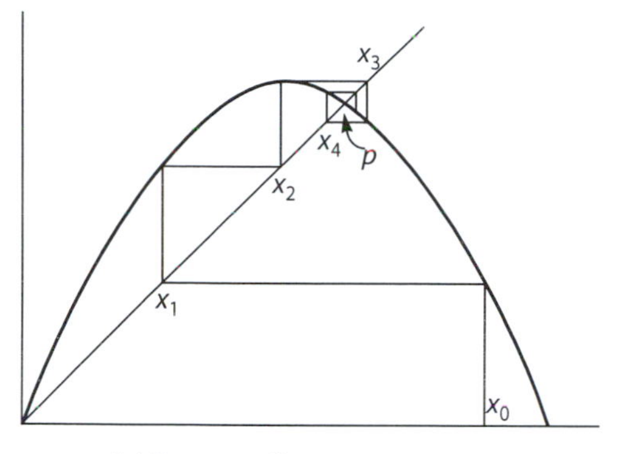

(a) $2 < \mu < 3$, convergence to p

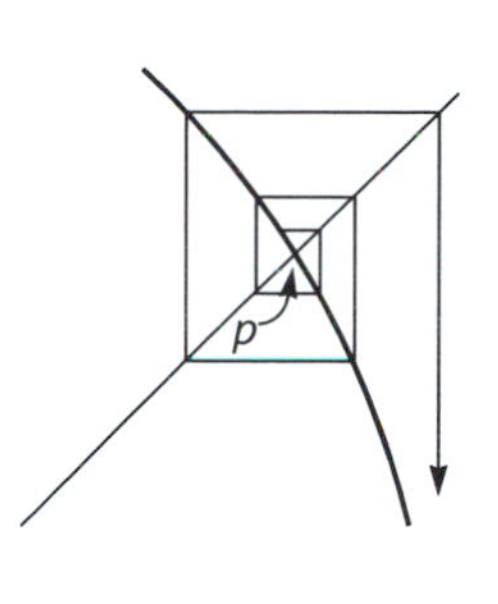

(b) $\mu > 3$, divergence from p

Figure 6.8 Graphical analysis of iterates of f_μ

diagram is constructed as follows. Start from some initial x_0. Go vertically to hit the curve $y = F_\mu(x) = \mu x(1 - x)$: the value of y at the intersection is to be x_1, the next input value for x. So if we now track horizontally to the diagonal $y = x$, the x-value at the point we hit will be the required x_1. Now go vertically back to the graph, to find the y-value $F_\mu(x_1) = x_2$; track horizontally back to the diagonal to intersect where $x = x_2$, the next input value. Keep on going.

For $2 < \mu < 3$, the resulting web diagram will look like Figure 6.8a: i.e., for any $x_0 \neq 0$, the web trail will spiral in towards the fixed point at p. But as μ increases, the maximum value of $F_\mu(x)$ also increases, and the slope of the tangent at p gets correspondingly steeper. And when $|dF_\mu(x)/dx|_p > 1$, as in Figure 6.8b, a web diagram starting near p will spiral outwards, *away* from the fixed point. In other words, p has become a repelling point.

It is easy to see, then, why p will cease to attract as μ increases. But where does the period 2 orbit come from? Well, suppose $F_\mu(q_1) = q_2$ and $F_\mu(q_2) = q_1$: then $F_\mu(F_\mu(q_1)) = q_1$, and similarly for q_2. So the period two points for F_μ are *fixed* points for the function $F_\mu{}^2(x) = F_\mu(F_\mu(x))$. Let's consider, then, what happens to *this* function – sketched in Figure 6.9 – as μ varies. Below the bifurcation point $\mu = 3$, $F_\mu{}^2$ intersects the diagonal $y = x$ just at the origin and at p (which must be a fixed point for $F_\mu{}^2$ if it is one for F_μ). And the iterates of $F_\mu{}^2$ converge to p: see Figure 6.9a. But as μ increases, the slope at p increases until orbits starting near p cease to be attracted, as the web diagram Figure 6.9b shows. However, the diagonal now intersects the curve at two new points q_1 and q_2 – so here are the two additional fixed points for $F_\mu{}^2$ – and *these* are both attracting. These new fixed points for $F_\mu{}^2$ constitute, then, an attracting period 2 orbit for the original F_μ. And so we get our period-doubling bifurcation.

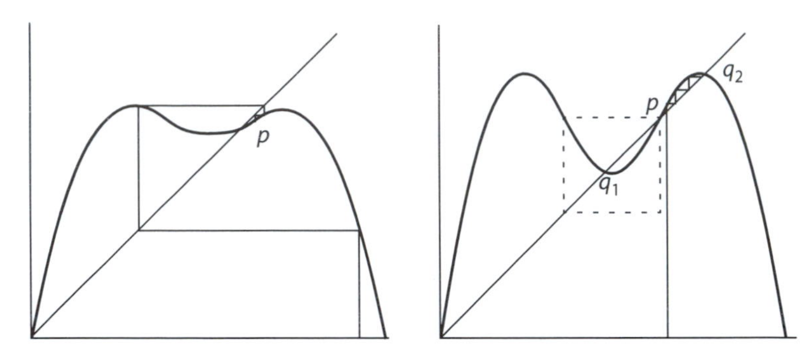

(a) $\mu < 3$, convergence to p (b) $\mu > 3$, convergence to q_1, q_2

Figure 6.9 Graphical analysis of iterates of $F_\mu{}^2$

And now note the key point that nothing in this explanation depends on the *exact* shape of the graph for F_μ. Suppose f_μ can, after rescaling (and perhaps truncation to its interesting region), be treated as a map of the unit interval, with $f_\mu(0) = f_\mu(1) = 0$, reasonably nicely behaved with a single maximum – so that f_μ is in effect a mild distortion of the logistic map F_μ. And suppose that the effect of varying the parameter μ is to change the slope at the fixed point where the graph intersects the diagonal. Then the same considerations show that the fixed point should again become unstable by a period-doubling bifurcation as μ alters.

Next, to see why the first period-doubling bifurcation should initiate a whole cascade, note that in the dotted region of the graph in Figure 6.9b, we get something very like a rotated, reduced version of Figure 6.8a. Hence we may well expect that, as μ increases further, the change in behaviour of iterates of $F_\mu{}^2$ near q_1 will recapitulate the change in behaviour of iterates of F_μ near p (for somewhat smaller μ). In particular, q_1 will become unstable and simultaneously new fixed points for $F_\mu{}^4$ will be born – so there will be another period-doubling.

The observation that, in a critical area, the graph of $F_\mu{}^2$ can be treated as an approximate scale version of F_μ, itself iterates. Thus part of $F_\mu{}^4$ is an approximate scale version $F_\mu{}^2$ (and hence part of that part is a smaller scale version of F_μ). So the change in dynamics of $F_\mu{}^4$ near a period 4 point for F_μ will again recapitulate what happens to F_μ – as μ increases there will be another period-doubling. And so on.

The same will hold for any f_μ which is a reasonable unimodal function. Further, the scaling relation that holds between small parts of the graphs of $f_\mu{}^{2n}$ and $f_\mu{}^{2(n+1)}$ is, in the limit, independent both of n and also of the exact shape of f_μ. The second point holds because in the end the scaling relation depends only on the shape of the graph f_μ just around its maximum: so if a unimodal map is like the logistic map in being – at least approximately – quadratic round its maximum and linear in μ, then we will encounter the same scaling behaviour in the limit. And, if we can somehow connect scaling behaviour with bifurcation rates, then the fact that there is this common scaling behaviour in the limit can be called on to establish that there is a common limiting bifurcation rate – i.e. to show that for unimodal maps with quadratic maxima, the rate tends to a constant δ. This connection *can* be made, though it is a decidedly difficult matter to establish formally.

Returning to the logistic map, it is worth briefly amplifying one final point. It was said that when $\mu = c_\infty$ (at the limit point of the series of period-doubling bifurcations), the behaviour of orbits is not chaotic, though it becomes so for $\mu > c_\infty$. How do we determine that? Well, consider a pair of orbits initiated by close points x_0 and x'_0, where $|x_0 - x'_0| = \epsilon$. These points are mapped to points x_1 and x'_1 such that $|x_1 - x'_1| \approx \Lambda(x_0)\epsilon$ (where $\Lambda(u)$ is the slope of F_μ at u, i.e. $|dF_\mu(x)/dx|_u$ – see Figure 6.10). If we follow the orbit one more step, we get to points x_2 and x'_2 where $|x_2 - x'_2| \approx \Lambda(x_1)|x_1 - x'_1| \approx \Lambda(x_0)\Lambda(x_1)\epsilon$. And the distance between the n-th iterates of x_0 and x'_0 will be (approximately) $\Lambda(x_0).\Lambda(x_1).\Lambda(x_2).\ldots.\Lambda(x_{n-1})\epsilon$. Now, as n increases, this factor becomes independent of the choice of x_0: and we can define

$$\Lambda = \lim_{n \to \infty} (\Lambda(x_0).\Lambda(x_1).\Lambda(x_2).\ldots.\Lambda(x_{n-1}))^{1/n}$$

which gives the characteristic growth/shrinkage factor per iteration of F_μ. If $\Lambda > 1$, then orbits that start close together will tend to spread apart, and so we have sensitive dependence. But in fact, for $\mu = c_\infty$, $\Lambda = 1$, and so orbits wandering over the attracting Cantor set are not truly chaotic. Still, as soon as μ is even slightly greater than c_∞, $\Lambda > 1$, and so then we do have sensitive dependence. (Note: It is more common to consider not Λ but the so-called 'Liapunov exponent' λ defined as log Λ. In terms of this, for close initial states, $|x_n - x'_n| \approx |x_0 - x'_0|e^{\lambda n}$ – compare (EXP), §1.5. A positive value for λ means sensitive dependence.)

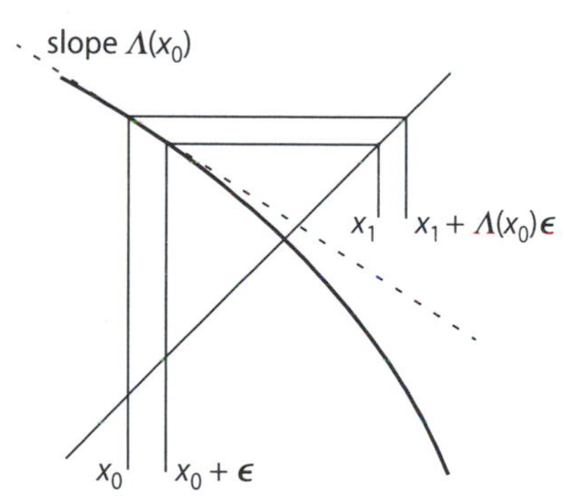

Figure 6.10 Behaviour of close orbits

6.5 The logistic map has considerable interest in its own right, and not just as a mathematical object: it is e.g. a possible model of the dynamics for the population level of a species in an environment with limited food (describing the way that the population level x_{n+1} in year $n+1$ depends on the level x_n in year n, with μ indexing the breeding rate). However, we motivated looking at the behaviour of one-dimensional discrete maps by noting that a trajectory in a many-dimensional continuous system may in some cases hit a suitably chosen 'surface of section' in what can be treated as one-dimensional set of points, so that key features of the many-dimensional continuous dynamics get reflected in a one-dimensional discrete dynamics on that set. We now return to exploring that connection.

We noted (Figures 3.2, 6.2 and associated text) that where there is simple stretching and folding back of trajectories, a one-hump, unimodal, map may arise very naturally in describing what happens at a suitable surface of section cutting across the trajectory bundle. But suppose that varying a certain parameter in the original dynamical equations has the effect of changing how much trajectories are spread apart before being folded back: the result then may be to vary the 'height' k of the related unimodal map – see Figure 6.11, and compare Figure 6.2. In this sort of case, then, we will expect that changes in the behaviour of the dynamics of the original continuous system (as the relevant parameter varies) will be reflected in the changes in the dynamics of a related family of unimodal maps (as *its* parameter varies). Hence, in particular, we may expect to find the same period-doubling route to chaos, and also expect to find the same universal

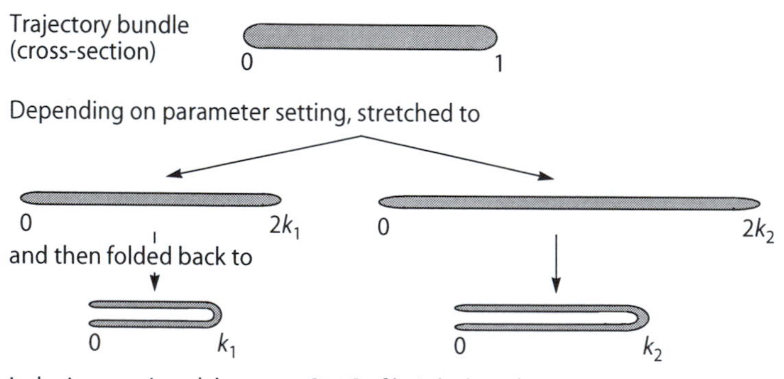

Figure 6.11 Possible effect of parameter changes

quantitative features such as bifurcation rates tending to the Feigenbaum constant δ.

And this is what we often *do* find. Consider again the Lorenz model (L*) with parameter r, given in §4.5. We noted (in the interlude on 'Bifurcations in the Lorenz system'; see especially Figure 4.4) that for $r \approx 100$ there is an attracting periodic trajectory, with three loops. Imagine a suitable surface of section P intersecting one of the loops. Then initially the trajectory will hit this surface at a single fixed point. But as r gets smaller, there is a first bifurcation; the three-loop trajectory becomes unstable, and it is supplanted by a six-loop attractor. This new attractor intersects the surface of section P twice; so a trajectory winding around it makes successive hits on the surface at two alternating points. Then as r reduces again, we get another bifurcation; hits on P by the new attracting trajectory will form an orbit of period four. And so on. The first few values of r where we get period-doubling bifurcations can be found, at least approximately, by computer trials; and the gaps between them nicely form an approximately geometric sequence, with their ratios of order δ.

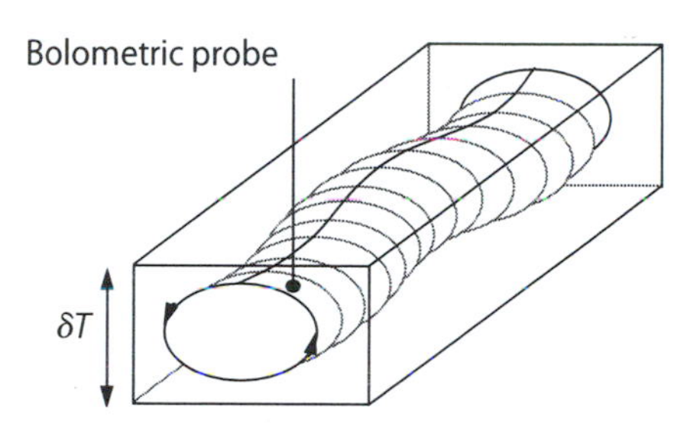

Figure 6.12 Convection roll with longitudinal wave

Computer explorations of other continuous models reveal that this kind of pattern is widespread; and we find the same when we turn from computer trials on mathematical models to empirical investigations of various worldly phenomena.

Take first the case of Rayleigh-Bénard convection flow in small cells of liquid where there is an imposed temperature differential δT between the top and the bottom of the cells. (This, recall, is the case which Lorenz was aiming to model – see the Chapter 1 interlude, 'Deriving the Lorenz equations'.) For small δT, steady convection rolls are set up, and the velocity and temperature at a given point will be constant. But as the temperature differential increases, the initially steady rolls begin to 'wobble', i.e. a wave starts running longitudinally along them (Figure 6.12). This will mean that, measured at a fixed point in the liquid, we no longer find a constant circulatory velocity, and likewise the temperature oscillates as the wave disturbs the liquid.

The easier quantity to measure here is the temperature at a point (we just insert a sensitive bolometric probe). When this is done, it is found that the temperature initially cycles periodically with some period φ. But as δT increases, the pattern of temperature changes suddenly gets more complex and the pattern of temperature oscillations now only repeats after 2φ. And then as δT increases again, there is another bifurcation, and the cycling period becomes 4φ. And so on. It is very difficult experimentally to control more than the first four period-doubling bifurcations (since they start to arrive so fast). But in the classic experiment using small cells of mercury, the rate at which the first four appear – i.e. the ratio between the jumps in the value of δT between successive bifurcations – was measured at 4.4 ± 0.1. Compare the Feigenbaum constant $\delta \approx 4.669$. Rayleigh-Bénard experiments using other liquids such as water and liquid helium have produced comparable values such as 4.3 ± 0.8 (note that, the rate for e.g. the first couple of bifurcations in the logistic map is approximately 4.27, so the agreement is excellent).

Analogous experiments have been run on other quite different physical systems which also show cyclical behaviour with doubling in the period of cycles as some control parameter varies. We get similar results across a variety of cases. To mention just two more examples: there are driven electrical circuits where the pattern in the oscillating voltage across a diode in the circuit shows period-doubling as the driving voltage is increased: the rate for the first five bifurcations has been measured at 4.3 ± 0.1. Likewise, there have been investigations using oscillating gas lasers which exhibit period-doubling at bifurcation rates of the same order. (For these and other experimental results, see e.g. the papers collected in Part 2 of Cvitanović 1989.)

6.6 It is all too easy to tell the story here in such a way as to engender a sense of deep mystery.

Start by noting that a variety of quite disparate physical systems exhibit period-doubling behaviour as an appropriate control parameter is varied, with the bifurcation rates being approximately equal. This universality is very puzzling. Then pluck out of the blue the logistic map, and note that – lo and behold! – there is period-doubling again, at the *same* kind of rate. Which is even more puzzling. For what has e.g. Rayleigh-Bénard flow or gas lasers to do with the logistic map? It can seem that there must be potent numerical laws of a previously unimagined kind at work in the universe (or so one might suppose, reading some popular presentations).

We have now seen enough, however, to understand that there is no real mystery here. The features of the logistic map that are in question – the presence of period-doubling and the rate at which the bifurcations pile up – are universal features of a wide class of unimodal maps. Which immediately turns the question 'What on earth has the logistic map, of all things, got to do with the Rayleigh-Bénard flow (and the other physical phenomena)?' into the rather more general question 'What has a unimodal map of some sort or other got to do with e.g. that kind of flow?' But there is no mystery about the idea that a dynamics may often be usefully modelled by considering the behaviour of evolving trajectories in a suitable phase space. So the question becomes: 'What has a discrete unimodal map to do with the behaviour of continuous phase space trajectories in competent models of Rayleigh-Bénard flow and other physical phenomena?' But to make the connection between discrete maps and continuous dynamics, all we need is the simple idea (§1 again) of looking at how a trajectory bundle hits a Poincaré surface of section cutting across the bundle. Which turns the question into 'Why do trajectory bundles in models of various physical phenomena often hit a suitable surface of section in a way that can be described using unimodal maps?' We then note that in energy-dissipating systems (§1.5 interlude) the dynamics doesn't preserve phase space volumes, so trajectory bundles get scrunched together. So we are left with the core question, 'Why should a dynamics which compresses trajectory bundles often result in a sequence of hits on a suitable surface of section which can be described using a unimodal map?'

And, as we have seen, a story about how trajectories may get stretched and compressed into thin sheets which are then folded back on themselves (to give confinement) provides the needed final link, at least schematically.

Applied to a particular case, this story may as yet have to remain schematic at the key final stage. Suppose we observe period-doubling behaviour at something like the standard bifurcation rate as a control parameter is varied. If we also observe the kind of overall order which suggests that trajectories must be confined, plus the apparent sensitive dependence which suggests that trajectories must peel apart, then we know that a model with stretching and folding looks a good bet. And we also know that for a wide class of such models, we can get period-doubling behaviour. Which gives us a partial explanation of the period-doubling behaviour. Only a partial explanation, to be sure (until we have a tolerably well-motivated choice of a more specific model): but there need be no deep puzzle about the *sort* of materials that will be needed to complete the explanation.

Intermittency

It is worth briefly describing one more 'route to chaos' to put alongside the period-doubling cascade and the birth of a chaotic attractor in a homoclinic explosion. And as with the period-doubling route, there are again universal features of this route which are independent of the fine details of the dynamics.

Consider again a family of discrete one-dimensional maps f_μ (with the f_μ linear in the parameter μ). And this time, suppose that in some region, the effect on the graph of the function as μ varies through a critical value μ_c is as shown in Figure 6.12a. Thus, on one side of the critical value, f_μ intersects the diagonal at a couple of fixed points q_1 and q_2, one of which is attracting. As μ passes through the critical value these fixed points first coalesce at q (where f_{μ_c} is tangent to the diagonal) and then vanish. But while μ is still close to μ_c, the old fixed point at q still exerts an influence, so to speak. As the web diagram in Figure 6.12b indicates, an orbit may be held up a long time in the neighbourhood of q – though it will slowly drift through the 'channel' between the curve and the diagonal, eventually emerging and then wandering away. Depending on what happens to f_μ elsewhere, an orbit may well eventually have to return to the neighbourhood of q and be held up there again (with the length of the stay in this region depending on the exact way the orbit is re-injected into the channel). In such a case, what we will typically

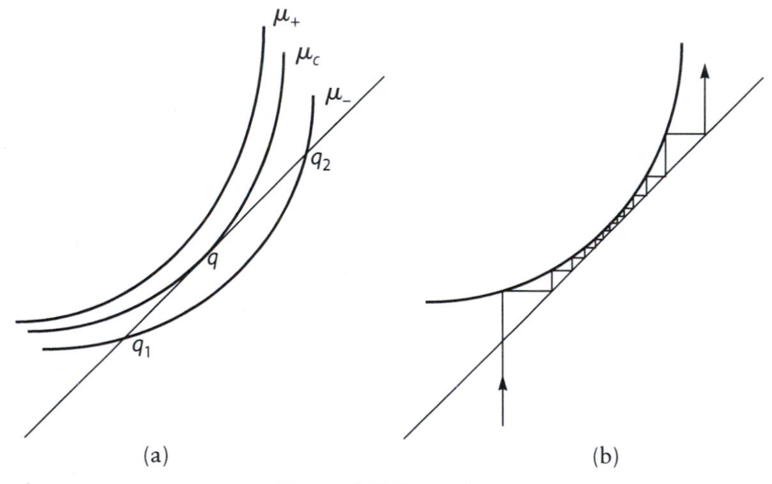

Figure 6.12 Intermittency

see – qualitatively – are episodes of 'laminar', almost steady, behaviour, interrupted at irregular intervals by bursts of much more varied behaviour as the orbit escapes and wanders away from q. An attracting fixed point has given way, by way of a so-called 'tangent bifurcation', to an *intermittency* regime.

Four quick observations. First, as μ moves further from the critical value μ_c the average period of time spent near q will get smaller – and it can be shown that, if the tangency at q is quadratic, then the average duration of the laminar phase will be proportional to $|\mu - \mu_c|^{-1/2}$. This scaling behaviour as μ varies is again a universal feature, i.e. it is independent of the exact shape of f_μ.

Second, there are a variety of empirical phenomena where we find periods of almost-regular behaviour interrupted at seemingly random intervals by bursts of much less regular behaviour. A dramatic example is the Belousov-Zhabotinskii chemical reaction. And although there are in fact other possible sources of intermittency, the experimental evidence in some cases is nicely consistent with the intermittency being the result of a quadratic tangency bifurcation with the predicted dependence of average laminar phase duration on the control parameter.

Third, suppose that orbits peel apart as they leave the vicinity of q by more than they later get squashed together when they revisit that neighbourhood (so there is a positive value for the Liapunov exponent). Then there will be sensitive dependence on initial conditions. There can also be periodic orbits – when a wandering orbit is re-injected into the channel near q at exactly the same place each time – mixed in with the more characteristic aperiodic orbits. So an intermittency regime may carry the signature of chaos again.

Fourth, and finally: consider the logistic map F_μ once more – or rather, consider the result of iterating it three times, the map $F_\mu{}^3$. Graphical analysis shows that $F_\mu{}^3$ undergoes a tangent bifurcation as μ decreases below the critical value $\mu_c \approx 3.829$. So for μ above this value $F_\mu{}^3$ has an attracting fixed point – this will be a period 3 point for the original F_μ and part of a stable period 3 orbit (and indeed, we are in the period 3 window visible in Figure 6.4). And below this value, we get chaotic intermittency behaviour. This explains why, to reverse the story, we have a chaotic regime for F_μ as we approach μ_c from below, and then out of the blue, after the critical point, a period 3 orbit is suddenly born.

6.7 We have described some of the intricate behaviour of the logistic map as its parameter varies (§3), and briefly noted that there is similar period-doubling behaviour, with the same quantitative features, in some physical phenomena (§5). There is no number magic here: §§1, 4 and 6 indicate the connections. The key ingredients to the connecting story are the universality results outlined in §4 – and it is in this area that some of the most intriguing of the new mathematics of chaos is to be found.

A final comment. We shouldn't underestimate the role of e.g. elegant universality results in fuelling interest in the study of dynamical systems among mathematicians. Various conjectures have been offered about why chaos theory should have become such a focus for research. Deep cultural explanations have been mooted; e.g. feminist philosophers of science have wanted to see a shift away from the hegemony of classic, masculinist, theories for controlling phenomena towards a recognition of the unpredictable and uncontrollable. But while such cultural explanations may have something to tell us about the wider reception of popularizing works on chaos, it is very doubtful that they are needed to explain the rapid spread of interest among mathematicians. As can be seen from interlude 'Period three implies chaos', proofs of results like the Li-Yorke theorem – or e.g. the beautiful ideas sketched in the concluding Chapter 4 interlude on homoclinic explosions – have all the good old-fashioned mathematical virtues of power, simplicity and elegance. What better reason for mathematical enthusiasm?

7

Explanation

7.1 In the last chapter, we saw how empirically observed period-doubling leading to apparent chaos might be explained (or at least, be partially explained). For we can show why period-doubling is endemic in a wide class of cases where other marks of chaos are also present. The relevant mathematics is intriguing; but on the face of it there seems nothing exceptional about the way that it may be brought to bear in explaining the phenomena. True enough, the modellings we are working with are often very partial and extremely idealized and perhaps (at this stage in the game) rather lacking in detailed physical motivation: but then, that's common enough in frontier science.

What goes for the explanations of period-doubling goes too, I claim, for other explanations in applied chaos theory. The new mathematical models may be distinctive; but the sort of explanations which their application yields is not. However, philosophers writing on chaos have repeatedly argued otherwise – they have claimed that there is something rather special about the explanations delivered by chaotic dynamical theories. Such explanations have variously been described as being characteristically 'ex post facto', 'qualitative', 'holistic' (or 'anti-reductionist'), and 'experimentalist'; it has also been suggested that we should recognize here a rather distinctive '*Q*-strategy' of explanation. But why so?

Suppose it is maintained, first, that prediction and standard modes of explanation go together – roughly, a standard explanation after the event deploys materials that, before the event, could have been used to frame a prediction. And suppose it is added, second, that applied chaotic dynamical models, because of sensitive dependence on initial conditions, are predictively crippled. Then it would follow that chaotic theories must either be explanatorily crippled as well or else be explanatory in non-standard ways. However, even if we accept the first premiss of this argument, which is certainly not compulsory, the second premiss is simply false as we saw in Chapter 4. Chaotic theories are richly predictive in quite straightforward ways. So the sketched argument gives us no reason to expect that chaotic dynamics, where explanatory, will differ in explanatory style from more familiar dynamical theories.

If that quick argument is rejected, why hold that chaos theory *is* special? In this chapter, we will consider and reject various claims about explanatory distinctiveness that have been mooted. But the point of this examination is not unremittingly negative. On the contrary, it will emerge that, while chaos theory may not be explanatorily *special*, it does vividly illustrate something of the explanatory style of much mathematical physics, a style not well captured in standard philosophers' portraits of science.

7.2 Have we set aside the 'quick argument' *too* readily? Compare the argument in Jesse Hobbs's paper 'Ex post facto explanations' (Hobbs 1993). He argues that in applied chaotic dynamics there can be no symmetry between explanation and prediction, so chaos theory is a domain where we find distinctively 'ex post facto' explanations. Thus consider the tumbling motion of Hyperion (one of Saturn's smaller moons). Why do the angular positions and momenta specifying its motion have the current values **v** (as opposed to some other set of values **v***)? The system, Hobbs writes, is

> presumably deterministic .... The catch is that the angular values cannot be predicted ... because Hyperion's motion is chaotic. That is, we can explain Hyperion's position and momentum at any time – there is nothing mysterious about what caused it or the laws governing it – but we can explain it only ex post facto. We have to see what the values are, first, being unable to anticipate or calculate them in advance except as one among a great many possibilities. (Hobbs 1993, 120)

This is an exaggeration. The values **v** *can*, at least to a reasonable approximation, be anticipated somewhat before the event. Hyperion is shaped like a planetary potato, and we can represent it by a suitable ellipsoid: it is not difficult to construct a dynamical model for such an orbiting ellipsoid and its tumbling motion. (The model will be a conservative Hamiltonian one, and so the complex motion here is not chaos induced by a strange attractor, but rather stochastic wandering over an equal-energy surface. But these details don't matter for the current philosophical point.) Knowledge of prior values can then be fed into the dynamical model to deliver at least short-term predictions in a perfectly standard way.

True, we will run into practical trouble applying the model over longer time scales because of error inflation. So even if we knew that the relevant values a year ago were approximately **u**, we couldn't use our model to explain why they are now **v** rather than **v***since – in the model – initial values observationally indistinguishable from **u** can lead, a year later, to more or less any **v*** permitted by energy considerations. But then our model which cannot give year-long predictions equally cannot deliver, ex post facto or otherwise, any differential explanation of why the values are now **v** rather than **v*** on the basis of the approximate initial values of a year ago.

Of course, assuming determinism (and the existence of precise values), it is true that the present values **v** are fixed by the *exact* values **u** a year ago, whatever they unknowably were. Perhaps we might count an appeal to this assumed fact as itself constituting a – highly attenuated – after-the-event explanation of **v**. Thus: 'Hyperion's state-variables currently take the particular values they do, not because of (say) the operation of random effects, but because Hyperion's state a year ago, whatever it was, determined the current values.' But note that the appeal to specifically chaotic dynamics has now dropped out of the picture – it is simply the underlying determinism which is doing any residual explanatory work.

The situation is this: where there is deterministic error-inflation, *chaotic or not*, there are limits on the prediction of future values of dynamical variables, and corresponding limits to the differential explanation of current values in terms of approximate initial conditions. And eventually, when time-spans are long enough, we may be reduced to offering attenuated ex post facto explanations of current values, appealing just to the presumed determinism. But equally, over short enough time-spans, again whether the dynamics is chaotic or not, we will be able to give perfectly standard explanations of why the variables take the values they do. So there can be no reason here to suppose that chaos theory, per se, distinctively 'gives rise to ex post facto explanations'.

Probabilistic explanation and explanation after the fact

Our dynamical model of Hyperion tells us (roughly) that the motion of an ellipsoid can be tumbling chaos. Likewise, to move on to another of Hobbs's examples, a partial model of atmospheric convection might tell us, for instance, that such-and-such atmospheric phenomena can exhibit periods of roughly regular behaviour interrupted by bursts of much more irregular behaviour at intervals distributed in such-and-such a statistical pattern (see the interlude 'Intermittency', §6.6). Thus, writes Hobbs, we might after the event explain unexpected rainfall in St Louis by noting that an area of instability had developed over Iowa. And what explains that instability? Circulation currents in the atmosphere can set up

> a positive feedback loop that exemplifies instability or chaos. It is typical of an outbreak of instability that a lot of things can happen as a result, or become more likely than they would be in the absence of chaos. In this case, rainfall is perhaps the most obvious – without instability there would have been no rain. Meteorological chaos thus was a contributing cause for the rainfall – necessary for it under the circumstances, and increasing its likelihood – thereby helping to explain it after the fact. (Hobbs 1993, 132)

But why only 'after the fact'?

Consider other cases of probabilistic explanation. *X* occurs, and we then explain it by noting that there was a chance of (say) 25% of *X* occurring, a chance that happened to be realized in this case. Such an explanation of *X*'s occurrence is not symmetrically related to a prediction, before the event, that *X will* occur – antecedently, all we could say is that there is a 25% chance of *X*. So, in a trite sense, probabilistic explanation is inevitably ex post facto.

In a little more detail: having observed a certain atom decay, we might explain the event after the fact by saying that it wasn't due to the bombardment of the atom by alpha particles, or to any other such external interference – it is just that the atom had a propensity spontaneously to decay within the observation period with a chance of 25%, and the propensity was actualized in this case. Paraphrasing a familiar treatment of probabilistic explanation due to Peter Railton (Railton 1978), we may further explain the decay by giving a nomologico-deductive account of the mechanism responsible for it, and by showing that our theory implies the existence of some physical chance that the mechanism will produce the explanandum in the given circumstances. But the core of this explanation can be equally put to work before the event; we may deduce from the theory that the chance of a certain atom decaying in the next minute is 25%. Exactly similarly with our explanation of the atmospheric irregularity. After the event, we may explain the development of the irregularity by giving a model of the mechanism responsible for it, and by showing e.g. that the model exhibits intermittency, which implies the existence of some chance that the system will yield irregularity. And the core of this explanation can again be put to work before the event, when we deduce from the model that the chance of an irregularity developing in the period of the weather forecast is (say) about 1%.

In sum, when chaotic theories deliver probabilistic explanations, then such explanations will in the trite sense be 'after the fact'. But again, there is nothing distinctive about the chaotic cases here (not all explanations in chaos theory are ex post facto; and when they are, it isn't the chaos that makes them so). So Hobbs's second case also fails to support the claim that there is a special character to explanation in chaotic dynamics.

7.3 It is very often said that chaotic dynamics, and indeed modern dynamical systems theory more generally, is characteristically qualitative rather than quantitative. Stephen Kellert, for example, emphasizes the

> distinction between specific *quantitative* predictions, the usual sort of which are impossible for chaotic systems, and *qualitative* predictions which are at the heart of dynamical systems theory. ... Qualitative understanding ... predicts properties of a system that will remain valid for very long times. It gives 'the general informations and the great classifications' by dealing with questions such as the periodicity and stability of orbits, the symmetries and asymptotic properties of behaviour, and the 'structure of the set of solutions'. (Kellert 1993, 101)

But we should not read too much significance into this observation.

For a start, to emphasize the point once more, a chaotic model can typically aim to capture a great deal more than purely qualitative features of the dynamics. It can deliver short-term quantitative predictions of 'the usual sort' (i.e. of the evolving values of dynamical variables). It can also, for example, give values for Liapunov exponents (like the λ in (EXP), §1.5) which measure the rates at which trajectories starting at neighbouring points diverge, yield predictions of numerical bifurcation rates as control parameters vary, and much more. On the other hand, old-style classical models always sought to do more than quantitatively match single trajectories. Just recall, to take the simplest example again, the standard high-school analysis of the motion of an idealized pendulum with *its* emphasis on the 'structure of the set of solutions' – i.e. on the way that period depends on length but not on amplitude, on the way that small perturbations knock the pendulum into a new stable periodic orbit, and so on. 'Qualitative understanding', in Kellert's loose sense, has always been quite central to dynamics, even if in chaos studies it is somewhat more to the forefront.

Not that an intuitive 'qualitative'/'quantitative' distinction will sustain much weight in any case. (As an exercise, take an older textbook – such as Herbert Goldstein's canonical *Classical Mechanics* (1959) – and try deciding which parts should count as 'quantitative' and which parts 'qualitative'. You will rapidly lose any confidence that an informal distinction will do much useful work, even pre-chaos.) Still, we *can* impose a sharp formal distinction here, by putting just results about topologically invariant properties on one side of the divide, and everything else on the other (see §2.2 for the notion of topological invariance: there indeed is an abstract subdiscipline labelled *topological dynamics*). Perhaps this sharp distinction is what writers like Kellert have in mind. But the trouble now is that, if we really narrow down the idea of the 'qualitative' in this way, it becomes even more clearly false that chaotic dynamics is essentially qualitative. To be sure, some interesting dynamical properties are indeed purely topological: for example, the basic distinction between periodic and aperiodic motion – the distinction between trajectories which are closed loops and those which are not – is preserved under deformation of trajectory bundles. But all the numerical properties of quite central concern to applied chaos theory – e.g. the size of Liapunov exponents, the fractal dimensions of strange attractors, the indexes of bifurcation rates – are not topological invariants, and so in our sharpened sense count as *non*-qualitative.

Geometric and topological ideas of a kind that used to be confined to the advanced study of dynamics have now percolated down even into elementary presentations of chaos theory, as indeed is already illustrated in this book. So against a limited background of high school dynamics it might seem as if there has been a 'qualitative' turn in the very general sense of a geometrical reconceptualization of dynamics which is new and specific to chaos theory. But such appearances mislead. In fact, any putative distinction between 'qualitative' geometric/topological explanations and 'quantitative' analytical treatments simply cross-cuts the distinction between chaotic and more familiar dynamical models: both approaches have been used in advanced work for at least a century.

Anti-reductionism

It is also often claimed that chaos theory is anti-reductionist in spirit. Thus, in a much cited *Scientific American* contribution on 'Chaos' (Crutchfield et al. 1986, 56) we read:

> Chaos brings a new challenge to the reductionist view that a system can be understood by breaking it down and studying each piece.

Likewise Kellert suggests that chaos theory involves what he calls a 'holistic' mode of understanding, which involves opposition to

> microreductionism as a methodological injunction [which] asserts that it is always appropriate to seek to understand the behavior of a system by trying to determine the equations governing the interactions of its parts. The fruitfulness of chaos theory militates against this creed. (Kellert 1993, 90)

But what, more carefully, is (micro)reductionism? Is it really the suggestion that, faced with, say, the problem of accounting for the orbits of sun and planets under mutual gravitational attraction, it is 'appropriate' to look at the equations governing the interactions of the molecules which are the parts of the sun and the planets? If *that* is microreductionism, then (so understood) the doctrine is crazy, and we certainly don't need to appeal to chaos to challenge it.

More defensibly, the microreductionist's thought might be that we should be able to explain why various laws governing the macro-behaviour of a complex system hold as well as they do, given the underlying micro-facts and given in particular the laws governing the interactions of the system's micro-parts – for plausibly, on pain of countenancing over-determination, the macro-laws normally need to be seen as in some sense 'coming for free', given the micro-facts. For example, the microreductionist seeks to understand why the classical gas laws work as well as they do, given that gases are seething systems of atoms or molecules, interacting according to certain micro-laws. But chaos theory doesn't impugn microreductionism thus understood. On the contrary, ideas from the theory of non-linear dynamical systems can be used to help execute the microreductionist's explanatory program – for example, when exploring the defensibility of Boltzmann's classic assumption of continuous re-randomization used in deriving classical laws from statistical mechanics (see the §1.5 interlude on 'Dissipative vs. Hamiltonian dynamics').

Here's a quite different kind of example of the way that ideas from chaos theory can be recruited to help out a reductionist project (touched on in the same *Scientific American* article, which talks of how chaos 'provides natural systems with access to novelty'). If there is to be evolution by effectively random mutation and natural selection, then the mutations need to be random enough for a sizeable space of possibilities to get explored; if similar environmental triggers always produced the same mutation, then evolution would be even slower than it is. Now, one way to get the needed effective randomness in a more-or-less macro-deterministic framework would be to build some sensitive dependence into the reproductive process, so that minutely different circumstances can trigger different genetic mutations, with all the consequent advantages to the gene line. The downstream result of the underlying chaos (if that is what it is) may be the emergence of real ontological novelty – ultimately, new species. But again, there is plainly nothing in this neo-Darwinian picture of chaotic mutations and natural selection which counts against any plausible reductionist project for explaining this emergence. Rather, we are speculating about how a dash of chaos in the process might explain how the Darwinian mechanism can work fast enough even in a more-or-less deterministic framework.

Now, we may very well be struck, as Kellert is, by the contrast between (say) the exceedingly simple logistic equation on the one hand and the startling complexity of the dynamics it describes on the other. Simply knowing the basic equations governing a chaotic system may, by itself, leave us as yet quite in the dark about what kinds of complex behaviour are exhibited by the system, or about how the system will respond to changes in various parameters (cf. Chapter 6). But plainly it would be quite wrong to think of the way that complex behaviour arises unexpectedly out of simple equations as of itself counting against a plausible microreductionism. Putting it briskly: microreductionism, if it is anything, is a doctrine about the relation *between* levels, macro and micro, while the phenomenon of simple-equations-with-complex-behaviour already obtains *within* any particular level of description. And once we see that, even within a level, simple laws can generate staggeringly complex behaviours (quite unexpectedly, though in ways that we might come to understand on further analysis), the microreductionist's posit that macro-complexities can ultimately be understood as generated from simple laws governing simple parts at the micro level should look more, not less, plausible.

7.4 Chaos theory doesn't break the usual symmetries between explanation and prediction; like traditional dynamics, it involves a mix of the quantitative and qualitative – the most we can say is that the proportions in the mix are different – and, if anything, it can be an aid and comfort to the sensible microreductionist. So far, then, there is no reason to discern a novel style of explanation at work.

But what of the often-remarked reliance of chaos theorist on computer 'experiments'. Here is Stephen Kellert again:

> Chaos theory often bypasses deductive structure by making irreducible appeals to the results of computer simulations. The force of 'irreducible' here is that even in principle it would be

> impossible to deduce rigorously the character of the chaotic behaviour of a system from the simple equations which govern it. The difficulty of deriving rigorous results about the simple models studied by chaos theory is notorious. Even such an exemplar of chaos as the Lorenz system has never been strictly proven to exhibit sensitive dependence on initial conditions. In the face of this difficulty, researchers regularly turn to what they call 'numerical experiments', that is, the use of a computer to simulate the behaviour of an abstract dynamical system by numerically integrating the equations of motion. (Kellert 1993, 91–92)

Granted, chaos theorists have to rely largely on numerical integration of equations like those in the Lorenz system. But there is nothing new about that. At the engineering end of applied mathematics, so to speak, it has always been thus. Indeed it is a lot worse than Kellert explicitly says: we often rely on the numerical integration of *simplified* versions of the equations we really want, without the benefit of any a priori proof that the approximation methods will work. For instance, given the intractability of the Navier-Stokes equations, fluid dynamicists have *always* run (as it were, under the aegis of the master-model) various more manageable approximating models geared to deal with different specific types of flow. And which approximation to use in which case has in part been justified 'experimentally', by seeing which one delivers tolerable results when numerically integrated. So there have been numerical experiments all along (now much easier to do on computers). In short, while Lorenz's discovery that his hyper-radical simplification of the Navier-Stokes equation for two-dimensional convection gives us a chaotic model was novel, it was certainly not novel for its 'experimentalism'.

As for the thought that in the case of chaos it is *in principle* impossible – or at least often impossible – to establish rigorously the 'character' of the chaotic behaviour of a system, I simply see no reason for supposing that this is true. Kellert goes on to remark on the computational costs of calculating individual long-term orbits in a chaotic system – the costs being so high that the line is blurred between 'impossible to calculate in practice' and 'impossible to calculate in principle'. But there of course are other ways of establishing, more or less rigorously, the chaotic character of a system's behaviour than by calculating individual chaotic orbits (cf. the analysis of the period-doubling route to chaos which was touched on in the previous chapter, or of the birth of chaos in a homoclinic explosion as described in outline in the final interlude of Chapter 4). Why should there be a limit to the further discovery of such results?

Still, there is surely something to the general idea that a kind of 'experimentalism' (to stick to Kellert's term) is at work here. This is brought out more clearly by Adam Morton in his discussion of what he calls the *Q*-strategy of explanation which he sees as characteristically, though not uniquely, involved in applied chaos theory (Morton 1991).

What is the *Q*-strategy? It involves what Morton calls 'dimension splitting' and 'transition analogy'. The first of these relates to the fact that our models in canonical form are governed by equations which specify the rates of change dx_i/dt of some quantities (the state variables x_i), while telling us nothing about how some other quantities (the parameters μ_i in the equations) might change – the parameters are, so to say, controlled from outside the model. To construct a model in this style, therefore, requires splitting the physical quantities to be represented into those to be treated as state variables and those which can helpfully be regarded, at least *pro tempore,* as control parameters (the best way to split things in a particular case may be rather unobvious). However, 'dimension splitting' in this modest sense is quite characteristic of almost *all* mathematical modelling, whether or not the solutions of the constructed equations are chaotic – it is part of the familiar background of old-fangled text-book cases in classical mechanics (as when, for a trivial case, we treat the angular displacement of a pendulum as a state variable and the downward force on the bob as a control parameter). Hence the novelty in applied chaotic dynamics can hardly be pinned to this so-called 'splitting'.

So if there is novelty to be found, it must be in the use of so-called 'transition analogy', and it is this which is the core of the *Q*-strategy. And by 'transition analogy' Morton means exactly the sort of thing we began exploring in Chapter 6. Thus, pure mathematicians have provided very elegant investigations of some simple dynamical systems that exhibit chaotic behaviour: for example, as we noted, there is an immense amount of work now on the one-dimensional logistic map. In simple contexts like this, we can explore various 'routes to chaos', in particular (say) the period-doubling route. We then find that there are similarities with the route to apparent chaos in a number of physical processes. We get (say) period-doubling again, at roughly the same rate in various processes, roughly equal to that we find in the logistic map. This and similar transition analogies are extremely striking. In many cases we do not know exactly *why* they obtain. And as yet we have to settle which analogies to apply in which cases on the basis of experimental investigation. But still, appeal to these analogies seems explanatorily illuminating.

And this is typical of work on chaotic systems. Here's Morton's description. We

> [t]ake simple and easily understood systems – including imaginary systems – and work out the pattern of dependence of their attractors on their control parameters. Then look for these patterns in more complex systems. Very often they fit. The attractors and the relations between changes in the control parameters and changes in the set of attractors that hold for simple systems fit the data for complex systems. ... [It] is often not clear why transition analogy works ... But the general phenomenon is now well-established: By working with simple systems one can develop a zoo of attractors, transitions, and routes to chaos, into the cages of which the behaviour of complex systems fits. (Morton 1991, 101–2)

Thus, we might aim to construct a taxonomy of the principal routes to chaos, which we can then begin to deploy in gestures towards explanations. 'Why does this physical system show just such and such a pattern of stages in its development of apparent chaos?' – Ah! There are experimental signs that we are dealing with a period-doubling case, and *this* pattern of stages is just what must happen in a very wide class of cases which exhibit the period-doubling route to chaos.

So, in the present state of the art, there is typically a mix of abstract analysis and 'experimental' investigation. To develop our gestures towards explanatory connections, obviously we want to get our toy cases of chaos to link up with more realistic models of the physics (chemistry, or whatever). We look for moderately well-motivated models of the phenomena, no doubt aiming for an acceptable trade off between faithfulness to the physical complexities and ease of mathematical investigation. And then the name of the game is to get the pure mathematics and the physics to connect, working from both sides towards the middle. Thus, for example, we prove universality results showing that the period-doubling rate of the logistic map is constant across a much wider range of structures (so taking a big step towards reducing the sense of mystery about why we keep encountering the same doubling rate in nature).

We are not yet very good at getting the physics and mathematics to join up (more on this in Chapter 8: and cf. the point of §6.6, where we stressed that there is no deep puzzle about the *sort* of materials that will be needed to fill out an explanation by transition analogy – but of course getting the details to work out in practice is another matter entirely). So, as Morton notes, often we can't predict in advance which type of chaotic model will best fit which type of physical system. In *this* sense, we are indeed at an 'experimentalist' stage in many of our investigations. But this fact again has nothing specifically to do with chaos, and has everything to do with the fact that we are opportunistically choosing between models in the absence of a fully understood physical motivation – a common enough scenario in immature science.

This last point about immature science is important. Morton says that the use of 'transition analogy' is the crucial ingredient of a distinctive explanatory strategy characteristic of work on chaotic dynamics. But in fact the sort of interplay he describes, between a recipe book of simple mathematical models and the recognition of patterns in the physical phenomena, seems to be an entirely familiar scenario in mathematical physics. We start (shall we say, back in the heroic days of quantum mechanics) with some notable empirical patterns; we hit on some toy mathematical models that apparently mirror in a rough and ready way some of those patterns. Perhaps we are guided by physical intuitions, or (hoping for the best) we are adapting some old models that worked in vaguely analogous cases; or maybe we are just struck by the neatness of a model and proceed without much clue about why the model works even as well as it does. Faced with a collection of opportunistically constructed models, we may initially have no real idea of how to relate the various models to one overall story. So the name of the game is then to generalize, deepen and unify the mathematics, and to get it all to make better physical sense, with the needed mathematics and our physical understanding evolving under mutual pressure.

And in this respect, in the character of its development from hints and guesses and inspirational gestures towards a (still limited) degree of genuine explanatory power, work on chaology seems to be – *pace* Kellert and Morton – pretty much business as usual for a developing science. Morton's *Q*-strategy is a quotidian one, at least in frontier mathematical physics.

These last deflationary remarks are not philosophically empty. For turn the claim around: take it to be not that work on chaos is boringly analogous to what goes on elsewhere in physics, but rather that what goes on in physics (rather often) involves the same improvisatory, 'experimentalist' interplay between mathematical ideas and physical understanding that is so vividly exemplified in the case of current work on chaos. We can then ask: How well is this kind of interplay captured in standard philosophical portraits of mathematical physics and of the sorts of explanation we find in the domain? Kellert and Morton in effect answer 'Not well at all'. And I agree.

The trouble is that philosophers have too often thought in terms of examples from what we might call textbook physics; and once the material is regimented into textbook form, much of the real fun is over. In particular, the mutual accommodation of mathematical models to physical insights has already taken place, and we are presented with the upshot – a selection of decently simple, physically motivated mathematical models which are known to be tolerably well-behaved, at least as applied in a canonical (and often heavily idealized) set of sample cases. By concentrating on textbook mathematical physics, then, you are therefore bound to miss out on the crucial prior process by which mathematical models get developed and selected. You are bound to miss, for instance, the way in which purely mathematical considerations can get to play a large role in the design and choice of models for physical processes. Historical studies can help to get the picture right (the S-matrix programme in quantum field theory is a nice example). But the current ferment of work in chaotic dynamics can perhaps serve even better to illustrate the point, especially since most of the mathematics and the relevant physical ideas are pleasingly accessible.

So, to put a positive spin on the discussion of Kellert's remarks about 'experimentalism' and Morton's discussion of his so-called *Q*-strategy: what goes on in mathematical physics is potentially often much more like what they note is going on in contemporary chaos theory than philosophers' pictures of science capture. Still, it is one thing to say that work on chaos is a vivid exemplar of common strategies in mathematical physics which philosophers tend to overlook; it is something else to say that chaos theory has its own very distinctive style of dynamic understanding. The latter claim, to repeat, seems simply wrong.

7.5 We can now respond to a question that we raised at the very outset (§1.6). We noted that the paradigm Lorenz model is derived by throwing away all the higher-order terms from what is already a multiple simplification of the Navier-Stokes equation; and this is rather typical of the way that chaotic models are constructed. So what level of credence can we possibly give to the result, especially when the model itself tells us that very small changes can have unpredictable big effects?

Evidently we should not rely on such a model too closely. We can not trust it, for example, for very detailed predictions of the time evolution of the target physical system (e.g. Rayleigh-Bénard convective flow), even within the limited time-horizon for practicable predictions of phase-space trajectories already imposed by sensitive dependence on initial conditions. However, some features of the model – such as the patterns in the routes to chaos via period-doubling, or via a homoclinic explosion, as control parameters are varied – may be relatively *robust*, i.e. be features which are also shared by variant models where other perturbing terms are thrown in to make the defining equations somewhat more realistic. And we might be able to appeal to these more robust features to extract useful predictions about the kinds of behaviour and the kinds of transition to be found in the physical system. 'Universality' results – such as the fact that period-doublings in a wide class of cases show the same limiting bifurcation rates – establish that certain model features can be particularly robust. But other structural and quantitative features too might survive fairly stably across a family of dynamical equations. For example, computer explorations may reveal that adding suitable small perturbation terms to a set of dynamical equations does not destroy the existence of a chaotic attractor for a certain range of parameter values (but, say, just tweaks its fractal dimension). Such robust features, then, are the ones to take seriously.

As it happens, however, the Lorenz model turns out to be in some key respects robustly *false* to the phenomena in Rayleigh-Bénard flow – for example, as the key parameter indicating the temperature difference between the top and the bottom of the box of fluid is increased, we get period-doubling rather than the predicted homoclinic explosion (cf. §§1.4, 4.5–4.6, 6.5). So this model is in fact explanatory only in the more attenuated sense of revealing something rather general about how relatively simple stretchings and foldings of trajectory bundles can lead to chaos. But the point of principle remains: we might reasonably hope to extract physical information from features even of a radically simplifying model when those features are shared across a suitable family of models. To put it in a slogan: the robust features are the candidate explanatory features.

This point about robustness can usefully be pursued a little further, in a more general setting. We noted in §3.1 that, by the lights of our own best theories, quantities such as fluid circulation velocity, temperature, and so forth are coarse-grained and cannot have indefinitely precise real number values. But these are exactly the kinds of quantities whose time-evolutions are modelled by textbook dynamical theories, chaotic or otherwise. So, in the world there are fuzzy-valued quantities (whatever exactly that amounts to): on the other hand, our applied mathematical theories deal in perfectly determinate real numbers. Hence, in pursuing standard mathematical modelling, representing fluid velocities or whatever by infinitely precise real numbers, we are not just going beyond our epistemic reach – we are inevitably fictionalizing. Our fluid dynamical models (say) have surplus content, pretending that there is precision in the values of relevant physical quantities where there isn't. So what justifies our idealizing, mathematizing practice?

Simplicity considerations are part of the answer: and the discussion of Chapter 3 aimed to show how even chaotic models can count as appropriately simple. But there is more to be said.

Working fluid dynamicists (or other applied mathematicians) are surely not going to see a great problem here, and will take a pragmatic view. Their picture will be roughly this. We make a fictionalizing move when we represent coarse-grained physical quantities by real numbers. We then run our mathematical dynamical model from these precisified initial conditions, with precisified parameter settings, which will in turn deliver unrealistically precise predictions about the future values of various quantities. But that's acceptable so long as we defictionalize at the end of the process, by recognizing that the predictions do need to be fuzzified again (and a good model will itself reveal something about how much fuzzification is needed by showing us how errors propagate). So, agreed, the precise values in the model are fictions. But no matter. We can still construe many features of the precise model tolerably realistically. In particular, we can take seriously, for example, those fuzzified predictions that are robust in the sense of not depending (or at least, not depending too sensitively) upon the particular choice of initial precisifications, or upon the particular choice of parameter settings. And more generally, we can take seriously those features of what our model tells us are sufficiently conserved across precisifications.

So there is indeed fictionalizing here – but what is the alternative? Could we (for instance) use some kind of imprecise representations for fuzzy quantities, and then directly do a species of fuzzy mathematics on them, and hence avoid the dog-leg through the traditional precise applied mathematics? However, the very idea of genuinely fuzzy mathematics looks dubious (and what currently passes under titles like 'fuzzy set theory' famously deals in precise numerical modellings of 'vague' sets). And waiving that point, any such development will doubtless be nastily complicated. Moreover, we will presumably want to preserve the empirical successes of traditional modellings, so we will want our imagined fuzzy mathematics to be at least broadly conservative with respect to results obtained by using traditional applied mathematics and slightly fuzzifying the results. So in advance of being presented with some more worked-out proposals, it looks as though developing a fuzzy mathematics would probably be all pain and no gain. Why bother, then, when there is a perfectly good way of filtering out the 'ideal' from the 'real' when faced with the problem of surplus content due to coarse-graining? We can apply the maxim *trust the robust* – that is to say, only take as seriously representional what is reasonably stable as precisifications vary.

It is part of the remit of modern dynamical theory precisely to explore the respects in which its own models or families of models are or are not robust in the various desirable senses. Thus, there may or may not be stability with respect to choice of initial conditions (we get a failure of this kind of robustness in chaotic regimes, where there is sensitive dependence on initial conditions). There may or may not be stability with respect to parameter settings (we may get a failure of this kind of robustness around particular values, say near the accumulation point in the period doubling route to chaos, where very small changes in parameter have big effects on e.g. the periodicity of the dominant attractor). There may or may not be stability with respect to perturbations of the governing equations (either by noise or by the addition of new terms: the 'structure of the set of solutions' may turn out to be fragile, and destroyed by small perturbations).

Computer trials can give good indications of where stability results hold. Indeed, numerically integrating a set of equations on a computer with its limited-precision arithmetic and continual accumulation of round-off errors can only show up model features that are already tolerably robust with respect to a certain amount of low-level noise and the coarse-graining of quantities. So the working applied theorist's effective policy of only treating as physically real (some of) those features of models revealed by computer calculations neatly meshes with the a priori maxim about only trusting the robust.

To summarize. Models with surplus content work. The alternatives look likely to be gruesomely more complex to no predictive advantage. It is much simpler to keep the models clean and precise, and to acknowledge the fuzziness of the modelled physical quantities at the stage when the models are applied to the world. And we in effect ignore the models' surplus content when it comes to applications, treating the surplus as fictional by considering as genuinely representational only those aspects of the models which are sufficiently robust. This, I claim, is in fact the applied theorist's implicit position on the question of how we should extract information from models with the excess content that comes from precisifying fuzzy quantities. And the remarks earlier in the section (about how to extract information from models that not just precisify but may also involve further idealizations and simplifications) simply pursue the same line of thought in a natural direction. Sufficiently robust features of models will be candidates for empirical and explanatory significance.

Robustness and supervaluationism

If we want to give a more technical gloss to the ideas about robustness sketched above, it is natural to think in supervaluationist terms.

Supervaluationism is a familiar story about the semantics of vague predicates, based on the idea that our linguistic practice is beset by semantic indecision about what precise properties we intend to denote. For example, there are lots of nice precise boundaries that can be drawn on the colour spectrum: we are just semantically undecided exactly where to put e.g. a blue/green boundary. Hence our colour predicates are left vague. But what we say using them is true so long as it is 'supertrue', i.e. comes out true however exactly we sharpen things up at the boundaries, and is false if it comes out false on every permissible sharpening. And a predication of (say) 'blue' to a borderline case will be neither true nor false, for it comes out true on some ways of fixing a sharp blue-green border, and false on others.

Note that the usual supervaluationist story is about how vague linguistic representations relate to precisely-bounded properties. What we are after, for present purposes, is a story that actually runs the other way about, and tells us how precise mathematical representations relate to intrinsically coarse-grained properties. Still, that in itself does not count against our taking aboard the technical apparatus while somewhat reworking the metaphysical motivation. To keep things simple, just consider again the outlined treatment of coarse-graining: the supervaluationist gloss would run as follows. Given coarse-grained physical quantities, we can represent them by using precise values in a range of permissible ways; so the representation of an initial set of coarse quantities by a precise phase space point is thus non-unique. Our mathematical dynamical model can then be imagined as being run from each such permissible initial point. The robust truths about the time-evolution of the coarse-grained physical quantities, according to our applied model, will then be the supertruths, i.e. those propositions about outcomes that work out to be true according to the model whichever permissible precisified initial state we choose. Predictions of perfectly precise values will then not be supertrue; but predictions about (say) values as falling into some range may well be supertrue (if the range is generous enough; though the worse the error-inflation and the longer the prediction period, the greater will be the width of the required range). Such range predictions can be robust with respect to choice of precisification of initial state, and so may be taken seriously.

We will consider just one possible objection to this attractive proposal. Doesn't standard supervaluationism, at least in its present application, have a rather disastrous technical bug? For it allows existential quantifications to be supertrue even if no instantiation is supertrue. So take a proposition like [M] 'There is a real number x such that quantity q takes precisely the value x' where q is coarse-grained. Then, by hypothesis, it is false of each number x that it is the number (exactly) giving the value of q. But on the supervaluationist scheme, doesn't [M] come out supertrue? – for on each permitted precisification of q there *will* be a suitable number making [M] true, though there will be different numbers in different precisifications. So, supervaluationism yields exactly the result we surely don't want.

But not so. Imagine someone objecting to standard supervaluationism about e.g. colour predicates in a similar way: 'The theory makes "There is a sharply-bounded colour property b such that *blue* refers to b" count as (super)true because on each precisification of *blue* there is some precise b it refers to.' This is a bad confusion. Put it this way: the imagined objection confuses reference in the sense of what's assigned in a precisification in the semantic model with real reference. Similarly with the objection to the supervaluationist account of robustness. It confuses values assigned to quantities in a precisification with real values of those quantities. (Or put it this way. Supervaluationism gives us a rule for evaluating the output of applied models—outputs like 'if the temperature at t_0 is about q_0, then at t_1 it will be about q_1'. But propositions like [M] are not themselves part of the output of a dynamical model, but belong in the commentary on the application of models, so [M] should not be treated as in the scope of the supervaluationist account.)

7.6 In chaos theory, there is a certain stress on the 'qualitative', but that is not particularly novel. There is, at the current stage of development, much appeal to very over-simplified models with flimsy claims to physical justification; but that too is characteristic, at least of frontier physics. There is intriguing use of e.g. new universality results; but again, this doesn't betoken a radically new style of explanation. There is a general issue about how to extract explanatorily useful information from idealizing models; again, though, this is not a new problem, and the headline response – 'trust the robust' – is already implicit in the practice of working theorists.

So far, then, the message is that – in its explanatory style – chaos theory is very much business as usual. But the story isn't over: there is more about real-world explanations in the next chapter. And we will find that a striking feature of empirical work in chaotic dynamics (another, related strand in its 'experimentalism') is the way that key features of the dynamics can often be reconstructed from the observational data without the benefit of *any* motivating physical account. Does *this* kind of modelling raise new conceptual issues?

8

Worldly chaos

8.1 Let's take stock. We now have some idea of the kinds of behaviour in dynamical models – involving sensitive dependence on initial conditions, aperiodicity, and the like – that are thought of as typical of 'chaos'. We have seen that models exhibiting these kinds of behaviour can at least in principle be richly predictive and that we can coherently think of them as candidates for approximate truth. We have seen too how such models can throw some explanatory light on natural phenomena (even when the models are constructed by radical idealization). So most of the main general issues about the status of chaotic dynamical models that we raised back in §1.6 have turned out to be fairly easily resolved. One major issue remains – business for Chapter 9 – concerning the claim that deterministic chaos involves a kind of randomness. But leaving that aside, chaos theory so far looks to be conceptually in good order.

But what is the state of play empirically? As remarked in §§4.6 and 7.5, the Lorenz model, our paradigm example of a model with chaotic behaviour, is in fact empirically rather unsuccessful in its intended domain. It would be good to be able to note some more robust empirical successes for chaotic models. The story, however, is a mixed one.

8.2 To repeat, it is no surprise that the Lorenz model does not work very well as an account of Rayleigh-Bénard flow, given the raft of radical simplifications involved in its construction (see §1.4). On the other hand, it *would* be a surprise if real-world dynamical systems didn't sometimes reveal Lorenzian behaviour. For first, recall that in a system where there is energy dissipation, a corresponding bundle of phase space trajectories will get scrunched up together – more carefully, a ball of phase space points S will be mapped by the dynamics into a smaller-volume set $S(t)$ (see §1.5). The simplest case is where there is contraction in all directions. The next simplest is where the trajectories spread apart in one direction and are compressed together rather faster in the other directions (see Figure 3.2a). This will tend to produce what are effectively two-dimensional 'sheets' of trajectories (with the local direction of flow and the direction of spreading defining the two principal dimensions: see Figure 3.2b). If these sheets are to stay confined in a finite region of phase space – as energy considerations will typically require – then they will need to be somehow folded back on themselves, making what is effectively (no more than) a three-dimensional structure. That's why we can expect frequently to find a dissipative dynamics (even one that initially demands to be portrayed in a higher-dimensional phase space) being controlled by an attractor that lives, for all intents and purposes, in a 3D space.

And now, second, recall that a Lorenz-like structure is – despite first appearances – actually a fairly simple way of folding back sheets of

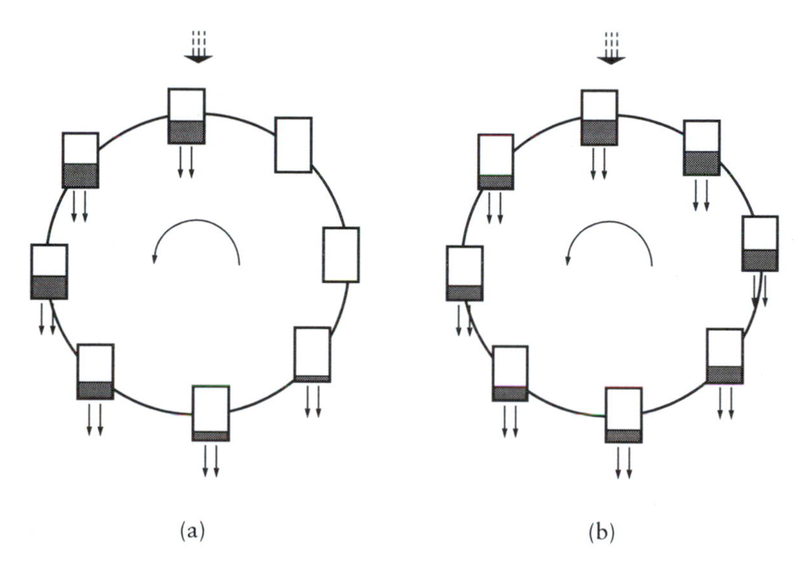

Figure 8.1 A chaotic water-wheel

trajectories in three dimensions (Figures 4.2, 4.3), and one which can emerge naturally out of changes in a still simpler dynamics (see the §4.6 interlude on 'Homoclinic explosions').

Putting these points together, we can predict that a Lorenz-like dynamics shouldn't – now we know what to look for – be too difficult to find. And it isn't. Here's one laboratory example, the *chaotic water-wheel*. The wheel is continuously fed from above but has leaky 'buckets'. We will expect that at low feed-rates (Figure 8.1a), the wheel will settle down to a steady rotation in one direction or the other, and the dynamics will be governed by a pair of point attractors – a pair because of the rotational symmetry. As the feed-rate is increased, however, the wheel will speed up. And eventually, as they pass more and more quickly under the feed, descending buckets won't have time to fill up enough to provide the wheel with sufficient torque to lift the ascending buckets over the top (those latter buckets were filled fuller, when the wheel was going a bit more slowly – Figure 8.1b). So the wheel will slow down and can reverse direction. Thus, with suitable rates of feeding and leakage, and an appropriate amount of friction on the wheel bearing, the point attractors are destroyed and we get varying speeds of rotation and erratic changes of direction. But now recall the description of the behaviour of the rotation-speed variable in the original Lorenz model, §1.4. We noted that this may first oscillate between various positive values (representing rotation in one direction), with the oscillations getting larger and larger until the value overshoots the origin and becomes negative. It then starts oscillating between various negative values until eventually being thrown back into a positive regime, with the lengths of these alternating positive and negative regimes varying unpredictably. The same pattern nicely fits the water-wheel: i.e. we empirically get a chaotic dynamics of exactly the kind that arises when a trajectory cycles around, being thrown from one wing to the other of a Lorenz attractor. And if we idealize the physical set-up (essentially by supposing that indefinitely many small 'buckets' are placed continuously round the rim), then we can go on to derive – without quite so many gross simplifications this time – equations equivalent to the general Lorenz equations again (see §4.5).

So, this is a toy case where we *do* plausibly find a worldly Lorenz-type route to chaotic behaviour (glossing that claim in the cautious style recommended in §3.4 – i.e. it is not being claimed that the worldly dynamical quantities evolve in an infinitely intricate way, but that there is a stretch-and-fold pattern in the dynamics which, *if* executed precisely, *would* result in true chaos).

8.3 In one way, the water-wheel case is about as good as it gets. An admittedly very idealized physical story – but one making use of well-understood principles applied to a complex of well-understood phenomena – leads in reasonably short order to a low-dimensional set of equations which are known (from a combination of analytical and computer investigations) to describe chaotic regimes for certain parameter values; and we find a decent fit with the empirical behaviour. But usually (as already remarked in §7.4) things are messier.

Consider the Belousov-Zhabotinskii reaction (§3.1). Initially, this was observed in a closed system, a beaker containing a mixture of chemicals interacting so as to produce a long-lasting series of oscillations of irregular length (showing up as dramatic changes in the colour of the mixture). In this closed system, the reactions must eventually cease: but reactions of the same type can be kept going indefinitely in pumped reactors. In a standard experimental set-up, this involves four reactants – as it might be, $Ce_2(SO_4)_3$, $NaBrO_3$, $CH_2(COOH)_2$, and H_2SO_4 – steadily fed into a continuously stirred reactor with the reaction product steadily removed. A probe into the mixture measures the concentration of bromide ions $B(t)$. With the right input rates, the value of $B(t)$ oscillates apparently aperiodically; and it has been argued that this aperiodicity corresponds to chaotic motion on a strange attractor rather than being due e.g. to some underlying periodic behaviour being disturbed by imperfect experimental control. But how can we decide?

Proceeding 'from the top down', i.e. beginning from a theoretical account of the reaction and trying to extract a chaotic model, is problematic. The reaction involves at least twenty other chemical species, and is not fully understood. And if we start with our best shot at a skeletal set of interlocking equations governing the rates of change of chemical concentrations, one for each known species, we get an intractable set of twenty-something equations. To extract a useful model, we have effectively to ignore a lot of reactions as minor effects, and then make guesses at the numbers to be filled in for the reaction rates in the remaining equations. There are various schemes on the market for doing this (such as the seven-equation 'Oregonator model'), some of which do and some of which don't exemplify chaotic states, and there are no obvious theoretical grounds for choice.

If we want more confidently to discern chemical chaos in the BZ reaction, then we need somehow to proceed as well 'from the bottom up', i.e. to start from the observed data-series – e.g. the values of $B(t)$ – and somehow recover information about the attractor which is governing the dynamics. But how can *that* possibly be done?

8.4 We here need to make use of a powerful and initially surprising mathematical result, which is roughly as follows. Suppose we are dealing with a system of dynamical equations in state variables x_i, describing a model with attractor A. And suppose that there is some observable scalar dynamical quantity s which is a nice smooth function of some of the x_i (note, though, that in a typical non-linear system, the values of the x_i are all interrelated, so a function of any effectively carries information about all). As the values of the x_i evolve over time, we get changing values of $s(t)$; and although s is a scalar, we can construct a d-dimensional vector y, by putting

$$y(t) = (s(t), s(t + \tau), s(t + 2\tau), \ldots, s(t + [d-1]\tau)).$$

That is to say, the d components of this trumped-up vector at time t are values of s, taken at time-delays of τ starting from time t. Now imagine a point representing $y(t)$ tracing out a path in d-dimensional space as t varies, drawing some structure in this artificial space. As the values of the original state variable x_i get pulled in towards the attractor A, the components of $y(t)$ will get pulled in towards an attractor S in the constructed new space. And *if d is large enough and τ is well-chosen, the attractor S in the artificial space of time-delay coordinates will typically be similar to the original attractor A*. Moreover, 'similarity' is not just a matter of topological equivalence (so that a simple limit cycle in the original phase space is matched by a limit cycle in the new space, and a non-periodic strange attractor is matched by a non-periodic attractor, etc.). There will be important quantitative similarities too. For example, we can measure how fast nearby trajectories peel apart or are scrunched together by using Liapunov exponents which give the exponential rate of divergence/convergence. The chaotic cases are ones where exponents are positive for some dimensions – giving some stretching apart – and negative in other dimensions – giving compression. And the spectrum of Liapunov exponents that characterize the original attractor A will be matched by the new attractor S.

This result is almost magical, and it is easy to see why it is of the greatest significance for empirical research: it means that, given data about the time-evolution of some single dynamical quantity, we can hope to extract key information about the shape of the 'real' multi-dimensional dynamics of the system. For example, given the data about the time-evolution of $B(t)$ in the BZ experiment, we can aim to reconstruct the attractor for the system (see §5 below for more details).

Rather than attempt to explain the magic here (though the next interlude does a little in that direction) let's instead pause to illustrate a couple of basic points by considering a very elementary but suggestive example from non-chaotic dynamics, namely a system that settles down to being a simple harmonic oscillator. And we'll just consider motion on the attractor. So, first working down from the theory to the phase space representation, a harmonic oscillator is governed by equations in the canonical form

$$dx_1/dt = x_2$$
$$dx_2/dt = -k^2x_1$$

and, with suitable choice of axes, the point representing the state of the system at time t has coordinates ($\sin kt$, $k\cos kt$). The attractor in our imagined system is thus an ellipse in phase space. Suppose in this case that the observed quantity s is just the value of x_1 (the displacement of the oscillator); then what we observe is the projection of this 2D trajectory onto the line $x_2 = 0$.

Now consider just that observational data, i.e. the series of values $s(t)$. Note that, apart from the extremal values, the same s-value will sometimes be followed by greater values of s, sometimes by lesser values (as the system oscillates through that value from different directions). So looking just at these s-values, they do not evolve deterministically. Assuming that we *are* dealing with a deterministic system, we need to split apart the values in order to 'unfold' the true dynamics; we can do this by considering *pairs* of values $(y_1, y_2) = (s(t), s(t + \tau))$ – pairs which share the same first value may differ in the second value, depending on the direction of oscillation. If we graph the behaviour of these pairs we get results as in Figure 8.2. Note two crucial facts: (a) for

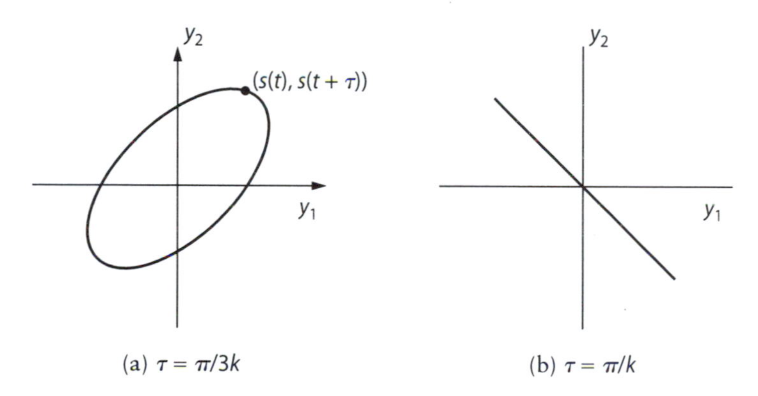

Figure 8.2 Time-delay reconstruction for oscillator

most values of τ we get a result topologically equivalent to the attractor in the original x_1-x_2 space – in fact, we get another ellipse – but (b) there are bad choices of τ where this doesn't happen, and we get no unfolding.

What if we had considered e.g. triplets $(s(t), s(t + \tau), s(t + 2\tau))$? With the right sort of choice for τ, the three-dimensional orbit of this point is again an ellipse – and likewise for higher-dimensional analogues.

Which illustrates another crucial fact: the first step(s) of unfolding the one-dimensional dynamics of $s(t)$ into a higher-dimensional space using time-delay coordinates will normally expose more dynamical structure (by separating cases where the same value of s corresponds to different points in the original phase space), but after enough steps, further unfolding makes no difference.

Embedding and phase space reconstruction

We concentrate here on qualitative, topological matters (quantitative issues about e.g. Liapunov exponents are beyond our scope). Some jargon: first, a function f is a *diffeomorphism* if it is (a) differentiable, (b) it is invertible (i.e. if $f(x) = f(x')$, then $x = x'$, so we can define the inverse function f^{-1}), and (c) its inverse is also differentiable. So, a diffeomorphism is a continuous one-one map which may e.g. stretch and distort regions but doesn't tear them, etc.; in short, it preserves topology. Second, a space X can be *embedded* into another Y if there is a diffeomorphism f such that $f(X) \subseteq Y$, i.e. X can be 'nicely' mapped into Y. Here's a simple example: a torus is a finite 2D space that can be thought of as a rectangle with its opposite edges identified as indicated by

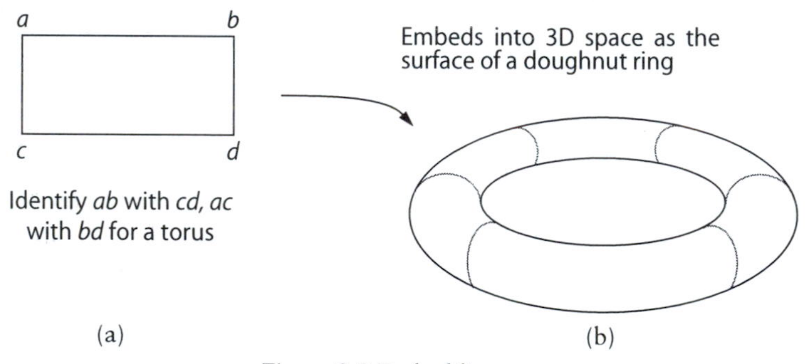

Figure 8.3 Embedding a torus

the labels in Figure 8.3a. But we usually portray a torus by means of an embedding into a Euclidean 3D space as in Figure 8.3b.

Suppose now we have a deterministic dynamical model described by equations in canonical form

$$dx_i/dt = F_i(x_1, x_2, \ldots, x_n) \qquad i = 1, \ldots, n$$

with the F_i nicely behaved functions. Suppose too that the 'observable' $s(t)$ is a nice smooth function of the state-variables x_i; and as before set

$$y(t) = (s(t), s(t+\tau), s(t+2\tau), \ldots, s(t+[d-1]\tau)).$$

Then y is a function of the state vector x; i.e. we have some equation

$$y = H(x)$$

(where H is compounded out of the 'nice' functions F and s). The crucial mathematical result is that, in the general case, we can choose d large enough so that for most τ

$$\text{if } H(x) = H(x'), \text{then } x = x',$$

(so that H has a nice inverse function H^{-1}) and is, moreover, a topology-preserving diffeomorphism. For this to hold, s has to carry enough information about the original values x_i: to take an extreme contrary case, suppose s were constant everywhere, so that $y(t)$ is a constant – inversion would be blocked. But leaving aside such singular cases, the requirements on s and F are modest – quite general conditions of smoothness suffice.

And when H *is* a diffeomorphism, the value of $y(0)$ at some time t_0 corresponds, via H^{-1}, to a unique value of $x(0)$; that value of x evolves over time according to the function F to a later value $x(t)$; and $x(t)$ corresponds, via H, to a unique value for $y(t)$. In other words, there is a derived deterministic dynamics for $y(0) \Rightarrow y(t)$, which is topologically equivalent to the original. In such a case, we say that H provides an embedding of the n-dimensional dynamics into the Euclidean d-dimensional delay-coordinate space.

A key question is how large d has to be in order to get an embedding. The answer, following from a general embedding theorem in topology, is that $d = 2n + 1$ is sufficient. But in fact, we are typically not interested in the whole n-dimensional dynamics, only in what happens on the original k-dimensional attractor (where k, the fractal dimension, will usually be much less than n): and to get an embedding that preserves the topology of the attractor, $d \geqslant 2k + 1$ will suffice. Indeed, we can often manage with less. But as an illustration of how we may e.g. need *three* delay coordinates to reconstruct a *one*-dimensional dynamics, take a case where the underlying dynamics is uniform motion on a circle, but suppose the observable $s(t)$ is some varying function of the angle θ. It could be that the 3D delay-coordinate reconstruction is a 'bent' figure as in Figure 8.4. Note that here the 2D

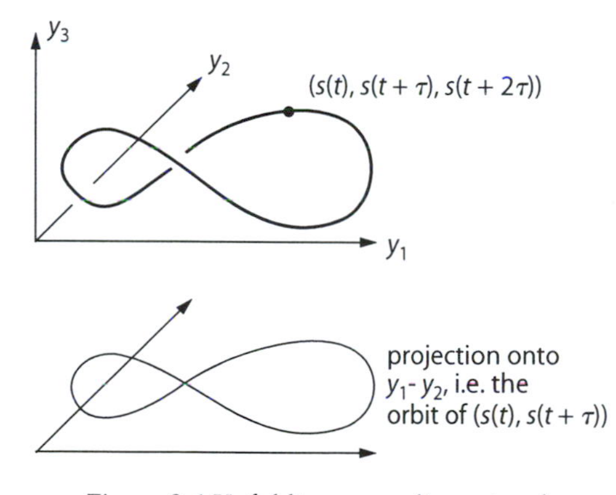

Figure 8.4 Unfolding a one-dimensional dynamics

projection onto y_1-y_2 has a crossing point, and hence *isn't* an embedding (the 2D motion of the point $(s(t), s(t + \tau))$ *isn't* topologically equivalent to the original dynamics).

Now reverse all this. Suppose we have observed the time-evolution of some quantity s; and we then construct corresponding d-dimensional vectors y^d, plot their time evolution and find (what looks like) an attractor A^d – and we do this for increasing values of d. Suppose (1) we find a value D such that the attractors A^d for $d > D$ are similar to A^D – there are various ways of checking for the requisite kind of similarity, by the likes of box-counting (cf. §2.3). So adding further dimensions after D makes no difference to what we find. And suppose (2) varying the value of τ involved in the construction of A^D also preserves similarity – that is to say, the general character of A^D isn't an artefact of a particular choice of time-delay. Then, *if* we have reason to believe that we are dealing with a dynamical system that can be thought of as governed by equations in canonical form with nicely behaved F_i, and reason to believe that s is a suitable smooth function of the state variables, *then* we have reason to suppose that A^D is an embedding of an attractor A for the original system. That is, A^D gives us key features of the original phase space dynamics.

8.5 Back to experimental work on the BZ reaction. In a classic paper, J. R. Roux et al. (1983) looked at the aperiodically oscillating time series for $B(t)$ – the concentration of bromide ions. Taking three time-delay coordinates with delay $\tau = 8.8$ seconds, they obtained an attractor portrait as in Figure 8.4a (compare the Rössler band, Figure 3.3). This is suggestive; but of course, the data-series for $B(t)$ consists in finite-resolution readings taken at discrete intervals. So how seriously can we take appearances here? Well, more is revealed by looking at what happens at a series of Poincaré surfaces of section cutting across the 'attractor'. These show that the bundle of trajectories is for the most part compressed into thin sheets which are indeed spread apart and then compressed together in the simple Rössler structure, confirming the appearance of Figure 8.5a. So the dynamics for $(B(t), B(t + \tau), B(t + 2\tau))$ *does* have just the characteristic stretch-and-fold signature of chaos. Further – as indicated in the last section – this chaotic attractor in the delay-coordinate space corresponds, on plausible assumptions, with the presence of a similar attractor in the underlying 'real' dynamics.

Moreover, suppose we examine the discrete 'recurrence map' $x_n \Rightarrow x_{n+1}$ produced by successive hits by a trajectory in delay-coordinate space as it intersects the indicated Poincaré section P in an almost-one-

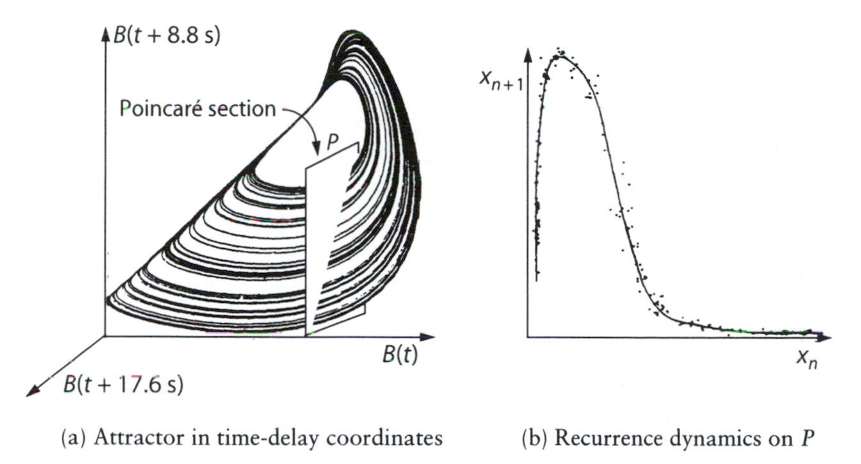

(a) Attractor in time-delay coordinates (b) Recurrence dynamics on P

Figure 8.5 Reconstructing the BZ dynamics

dimensional curve (see §6.1 for the general idea). The result is shown in Figure 8.5b – i.e. we find that the map is more or less a smooth unimodal one. But we are now familiar with the idea that the presence of unimodal, one-hump, maps is associated with the period-doubling route to chaos (as a parameter varies, changing the height of the hump). We might reasonably conjecture, then, that the chaotic attractor in Figure 8.5a might be reached by period-doubling as some appropriate control parameter governing the BZ reaction is varied, reflecting a corresponding period-doubling in the real dynamics. And indeed, such period-doublings in the chemical oscillations have been found experimentally (exhibiting moreover the standard 'universal' features of period-doubling in unimodal maps). So, putting it the other way about, we have a partial explanation of why there is period-doubling: it is due to the sort of stretch-and-fold dynamics that leads to unimodal maps.

Now, in the last chapter, we remarked on the 'experimentalist' character of work in chaotic dynamics: the to-and-fro between guiding theory and the construction of applicable mathematical models tends to be opportunistic and ad hoc, as when we choose among ways of simplifying the theory on the basis of what produces experimentally plausible models. But aren't we here involved with a more radical kind of experimentalism? We are not just choosing among candidate models (imperfectly justified by chemical theory) on the basis of experimental fit; rather, we are constructing a mathematical model bottom up, out of an experimental data-sequence, and then using facts about the model – facts derived from universality theory – to predict/explain other data, apparently without needing the background chemical theory at all.

But we shouldn't leap to supposing that some radically new explanatory style is in play here. For consider the situation more carefully. It is only because of our general background chemical understanding that we recognize the BZ reaction as apt for treatment as a high-dimensional dynamical system. And again, it is in virtue of our chemical understanding of such systems kept far from equilibrium that we know that this is a strongly dissipative system – so phase space volumes in the dynamical system will strongly contract. That gives us reason to expect the formation of low-dimensional attractors. And *that* (together with the considerations of the previous section) gives us reason to suppose that a low-dimensional attractor found when we use the time-delay construction is likely to correspond to the 'true' dynamics. So our background chemical understanding of the BZ reaction *is* after all involved in our justification for taking the empirically constructed attractor seriously. True, we do not need to appeal to a detailed theory of the chemical processes: we only need a rather general, attenuated, understanding. But then, we only get out of the construction an equally attenuated explanation of what is going on: 'there is the kind of stretch-and-fold dynamics that goes with period-doubling, and *that's* why we find together the features characteristic of a period-doubling route to chaos'.

8.6 There seems, then, to be pretty conclusive evidence that deterministic chaos can occur in chemical reactions. And the same holds across a wide range of other physical phenomena – e.g. in the behaviour of gas-lasers, driven electrical circuits, and electrically stimulated cells, as well as many further fluid dynamical examples to set alongside Rayleigh-Bénard flow. In each case, of course, the claim that there is worldly chaos must be interpreted with the now usual caveats: so, more carefully, we should say that there is, in certain regimes, a time-evolution of state variables that, in a phase space representation, yields a stretch-and-fold dynamics of a kind which, *if* developed perfectly precisely, *would* be truly chaotic. And so, employing the usual canons of allowable theoretical idealization of worldly dynamics, appropriate models for the relevant phenomena will be chaotic.

In a way, though, this is a modest claim, and it is actually quite compatible with the chaologist David Ruelle's rather pessimistic overall assessment of the empirical impact of chaos theory:

> The mathematical theory of differentiable dynamical systems has benefited from the influx of 'chaotic' ideas, and on the whole has not suffered from the current evolution. ... The physics of chaos, however, in spite of frequent triumphant announcements of 'novel' breakthroughs, has had a declining output of interesting discoveries. (Ruelle 1991, 72)

To be sure, it has been a major breakthrough to recognize that what initially looks like complex, noisy data (what looks to be chaotic in the everyday sense) may in fact be produced by a relatively simple non-linear dynamics acting to produce a 'strange' attractor (may be chaotic in the technical sense). Though, as we noted before, now we understand what to look for, it is no surprise that we find signs of empirical chaos in many domains: so totting up further empirical cases indeed has declining interest. What would be exciting would be to find more cases where we can get a detailed and physically plausible theory to output empirically competent chaotic models without relying too much on opportunistic and ad hoc simplifications, etc. Such cases are in fact slow in coming.

Chaos in the head?

It would be tedious to review all the various hints and suggestions that have been made by philosophers about the wider significance of the possibility of worldy chaos. This interlude looks at just one strand of ideas, and sounds a note of proper caution.

Our brains may well be among the empirical chaotic systems to be found in the world. The mathematical analysis of signal propagation in neural nets together with some suggestive experimental work indicates that certain states of the brain may well evolve with a non-linear, chaos-prone, dynamics. This isn't the place for an extensive treatment of the relevance of dynamical systems theory to theoretical psychology or the philosophy of mind. But occasionally, the possibility of chaos in the head is seized upon in order to draw large, over-hasty, conclusions.

Consider the following line of thought from the philosophy of mind – call it 'AM', as it is a version of 'anomalous monism'.

> The same psychological state can be physically realized in various ways; these different physical realizations can give rise to later physical states (according to physical laws); but these later physical states may realize *divergent* psychological states. So, while there may be strict laws at the physical level, there are no strict psychological laws.

There no doubt exist probabilistic psychological causal relations: wanting a beer right now, believing the beer is in the bar, and believing that the bar is down the corridor, does *tend* to get us down the corridor in the absence of defeating beliefs or desires. But the thought is that we don't have determinism at the psychological level, nor even determinism 'for all intents and purposes': qualitatively identical psychological-states (identical, that is, as far as our coarse-grained everyday folk theory is concerned) can – with significant probabilities – yield divergent outcomes (divergent even by the lights of the coarse folk). For an extreme example, when Buridan's Ass plumps for the left bale of hay while his duplicate twin plumps instead for the right bale, we can have the same folk-psychological antecedents, followed by different actions. Only the grip of dogma could make us insist that really, all along, there was a prior salient folk-psychological difference between the antecedent states of the twin asses, i.e. a difference which is discernable at the level of common-sense psychology. Or at least, so the argument goes.

Understood like this, AM *may* well be true. And the question whether AM *is* true matters, for example for some versions of the free-will problem. That's because the idea of *psychological* determinism is intuitively inimical to freedom (we think of ourselves, when in a Buridan's Ass situation: 'I freely plumped for the India Pale Ale; but I could, keeping fixed the preferences with which I approached the bar, have plumped for the Abbot's Ale'). And if – as AM implies – psychological determinism *doesn't* follow from any underlying physical determinism, then that removes *one* major reason for supposing that physical determinism is incompatible with freedom.

So far, though, this is routine and familiar. The novelty comes with the thought, encountered in discussion, that AM gains new support from appeal to the possible chaotic dynamics of neural nets. Suppose there is neural chaos: then, the line of thought runs,

1. A given initial type of psychological state can be realized in a variety of similar physical brain-states.
2. But (by hypothesis) the time-evolution of the brain-states in question is sensitively dependent on initial conditions – i.e. we may get markedly different *physical* upshots arising from very similar initial states.
3. Hence we can get (with significant probability) markedly different *psychological* upshots arising from the same initial psychological state.

Well, grant (1), and suppose (2) is true in addition. The trouble is that (3) just doesn't follow. Compare, for example, the following argument:

1*. A given initial type of thermodynamical state (e.g. a certain temperature) can be realized in a variety of physical states characterized by different position/momenta distributions of the particles in a gas.

2*. But the time-evolution of a state with a given position/momenta distribution is sensitively dependent on initial conditions – i.e. we may get markedly different distributional upshots arising from very similar initial states.

3*. Hence we can get (with significant probability) markedly different thermodynamical upshots arising from the same initial thermodynamical state.

This argument is plainly invalid, for the premisses are true and the conclusion false. Moreover, putting it crudely, we appeal to (2*), or at least facts like it, to explain *why* the likes of (3*) are false. Recall, the hypothesis of micro-chaos is an essential premiss in a Boltzmann-style derivation of the claim that classical gases approach equilibrium with overwhelmingly high probability.

Hence, we can't in general infer from underlying micro-chaos that there is significant macro-indeterminism with respect to higher-level properties. So the sketched form of argument is invalid. It could be that the variant physical states realizing a given mental state do yield very different physical upshots. But why shouldn't these very different physical upshots all in turn realize one and the same sort of psychological upshot? That, after all, is *precisely* what the familiar thesis of multiple realizability implies might happen.

The argument from (1) and (2) to (3) is perhaps tempting if you presuppose an over-simplistic model of the relationship between psychological states and physical states. Imagine that psychological properties just impose a simple *coarse-graining* on physical state-space; that is to say, pretend for a moment that a psychological predicate just (as it were) picks out the box in which the physical state is to be found. Then in a system with sensitive dependence, same-box starting points can lead to different-box outcomes; so we'll get the AM result that the same psychological state can have different psychological up-shots. But why should we accept the pretence that a simple coarse-graining even approximately represents the way that psychological predicates relate to physical states?

Certainly, some intriguing work on brain dynamics seems to count strongly against the idea. Walter Freeman and his associates have done studies on the reaction of a rabbit's olfactory bulb to presented stimuli – as it might be, carrots and garlic (Skarda and Freeman 1987). Inserted probes measuring electrical activity give oscillating readings looking much like dynamics on a strange attractor – with different stimuli producing activity settling down towards different attractors. Freeman clearly *rejects* the idea that psychological states correspond to simple neighbourhoods of an appropriate physical state-space; he suggests instead that some psychological states (olfactory perceptions) correspond to *large structures* in state space. We can't explore this fascinating suggestion here: but it is worth noting that Freeman's idea gives *no* support to AM. Here's the thumbnail argument. Suppose that what makes a current point in physical state-space correlate to a certain psychological state is the fact that it is *near the right strange attractor.* Then, of course, two neighbouring initial points near a strange attractor can yield trajectories which diverge from each other (that's what makes the attractor *strange*); but these resulting trajectories will stay near the attractor (that's what makes it an *attractor*). That will give us micro-chaos but macro-psychological stability. And the move from one macrodynamical state (defined by its attractors) to another as control parameters change can be as deterministic as you like.

The moral? The significance for psychology and the philosophy of mind of such chaos as there might be in the head will depend crucially on the details of the story about what features of brain-state, identified in what ways, correlate with psychological states. And at this stage in the game, it is very far from clear how the story will unfold.

9

Randomness

9.1 Return once more to our paradigm of a chaotic system, the Lorenz model. The time evolution of trajectories in this system is deterministic. But the precise sequence of events – so many turns around the left wing of the attractor, followed by so many turns around the right wing, followed by another visit to the left wing, and so on – typically seems quite patternless: the visits to the two wings appear to be randomly distributed. Is there any good sense, however, in which there really *is* objective randomness here? What are we to make e.g. of Joseph Ford's often-quoted remark 'chaos is merely a synonym for randomness' (Ford 1989, 350)? If there is a notion of randomness appropriate for describing deterministic chaotic *models*, can it usefully be carried over to describe the *world*? And how can we empirically distinguish true deterministic chaos and its sort of randomness from mere random noise? These are the central questions for this chapter.

But first, a brief comment about the claim that standard chaotic models are deterministic. John Earman's remarkable *Primer* (Earman 1986) teaches us that many questions about determinism are troublesome and can require considerable care to get straight. Still, the claim being made here does seem fairly unproblematic.

Earman himself starts, as is now popular, from a definition of determinism for *worlds*. A world w is said to be deterministic if for all $w^* \in W$ (where W is the set of all possible worlds which share the laws of w), if w and w^* agree at any time, they agree for all later times. There is an issue about how to spell out 'agree at a time' for complex space-time structures. And of course, the definition is only as clear as the notion of a 'law', which is to say it is not very clear at all. But waive those issues. The question is how to get from a definition of determinism for *worlds* to an account of determinism for *theories* (the notions are no doubt related, but in rather complex ways). Take, for example, the familiar issue whether Newtonian celestial mechanics, with its treatment of point masses moving under mutual gravitational attraction, is always deterministic. If the issue is approached in terms of possible worlds, the issue comes to something like this: are worlds where the laws are the Newtonian ones and the *only* forces are the gravitational ones deterministic? (The possible presence of *other*, randomly fluctuating forces, is surely not germane to the question of whether Newtonian celestial mechanics is a deterministic theory.) But then, what kind of 'worlds' can these be, where there are only gravitational forces? It would be stretching a point to think of these as possible worlds in the sense of 'collection[s] of possible events representing alternative possible histories to that of the actual world' (to borrow Earman's words). In fact, don't the supposed barely Newtonian worlds – from which everything has been removed apart from gravitational structure – sound pretty much like abstract *models* in more or less the standard model-theoretic sense? So, avoiding the detour through possible worlds, it would be better to say straight out that a theory is deterministic if any pair of models for the theory which agree at one time must agree at all later times. Which is Richard Montague's classic definition of determinism for theories (Montague 1974).

Note that the special emphasis on *laws* has now dropped out of the picture; this seems desirable (and not just because of the obscurity of the idea of a law). For example, we surely want to say that the classical theory of a vibrating string stretched between two points is deterministic. But that theory invokes not just background laws but also the special boundary condition that the displacement at the end-points remains zero at all times. Earman is famously pernickety about such boundary conditions concerning the future. Thus consider his claim that some Newtonian theories are non-deterministic because they impose no upper bounds on velocities, and hence they allow 'space invaders' (i.e. particles which – asymptotically from time t_1 – come in from infinity, perturbing values of the states variables at some later t_2 in a way that is therefore not determined by their values at the current time t_0). The obvious counter is to insist that the Newtonian theories for isolated systems assume null boundary conditions at infinity. Earman regards banning space invaders by setting the future boundary conditions at infinity as 'little more than a hypocritical refusal to consider the possibility of unpleasant surprises', and claims that it constitutes 'a departure from pure Laplacean determinism by requiring the specification of future data'. But while this point may have some force when determinism for *worlds* is in question, though that is debatable, it surely doesn't carry over to determinism for *theories*, when what we want to know is very often just whether laws *plus certain boundary conditions* determine unique outcomes.

The Montague definition of determinism, however, is really framed to fit the view of a theory as the deductive closure of some body of axioms. If, as has already been suggested (see §5.3), we should conceive of mathematized theories more along the lines of specifications of abstract structures plus rules of application, then the story can in principle get more complicated. For the core mathematical structure of a theory could have an intuitively deterministic dynamics, yet the intended application rules be probabilistic (allowing, say, for low-level random noise); and such a theory would presumably not be counted as deterministic overall. Still, in classical cases with straightforward application rules, it is entirely natural (and in keeping with the basic Montague/Earman picture) to deem a dynamical theory deterministic in its domain so long as the core mathematical structure is uniquely specified – up to isomorphism – and lacks singularities, incorporating a unique trajectory through the space of state-variables for each initial point in the domain. Standard existence and uniqueness theorems for solutions of linked families of differential equations immediately deliver the result that typical chaotic systems like the Lorenz model are indeed deterministic in this quite unproblematic sense (see the Chapter 1 interlude 'Dynamical systems and dynamical equations'). And for our purposes the matter can be left there.

9.2 It is no news that indeterministic processes can accidentally produce highly ordered products: make an indeterministic device for shuffling a deck of cards – e.g. build in a uranium source plus a geiger counter to make a random trigger – and it may yet produce a bridge hand of thirteen spades. It is no news either that deterministic processes can produce passably random products (i.e. highly disordered, patternless products): consider random number generators in computers. So, to put the point in headline terms, we need to distinguish *randomness of processes* from *randomness of products*. If there is to be randomness in chaotic models, it must be randomness in the product sense – since, by hypothesis, we are there dealing with models with thoroughly deterministic dynamics (the 'processes' are entirely non-random).

What, then, is it for the 'product' of a chaotic system to be random? Suppose, without too much loss of generality, that some aspect of its behaviour is coded as a binary sequence. For example, we might code the behaviour of trajectories near the Lorenz attractor by sequentially recording a '0' for a loop around the left wing, and a '1' for a loop around the right wing. Then the question is: what is it for a coding binary sequence S to be random in the required sense (i.e. patternless)?

Concentrate on the finite case: for we want to make sense of the claim that this or that finite empirical data-sequence produced by a real-world chaotic system is 'random' (though compare the final remark of §6 below). There is a familiar battery of distributional statistical tests for randomness in finite sequences. So why not just say that, for our purposes, *S* is patternless if it is *statistically random* (i.e. if the hypothesis that each digit in the sequence has a fifty/fifty chance of being a zero passes favoured statistical tests for the acceptability of such a hypothesis)? To illustrate, take the simplest sort of test: if *S* is truly patternless, then we expect that 0's and 1's will occur roughly as often; and in neighbouring pairs, we will find 00 and 11 pairs about as often as each other, and half as frequently as mixed pairs; and so on. We now need a measure of how far a bunch of observed distributions in *S* depart from what would be typical for a sequence produced e.g. by tossing a fair coin. The statistician's standard chi-square measure can be invoked here; and then one test of patternlessness in *S* will be a matter of yielding a 'reasonable' chi-square measure.

But we won't pursue the details further here, which do indeed get messy (see Knuth 1981, §3.3). For writers on chaos theory tend to work with a significantly different approach to patternlessness for finite sequences due to Andrei Kolmogorov, Gregory Chaitin and Ray Solomonov (henceforth KCS). I will be arguing that this approach is in fact less attractive than it initially appears. I go on to note, however, that chaos theorists' claims about deterministic product-randomness are in fact quite insensitive to the account of patternlessness that we give; so while the exact status of the popular KCS account is debatable, in the end this matters rather little for our particular purposes.

The leading idea of the KCS approach is elegantly simple. Compare for example the two finite sequences

11011001111101101011010101000010111010011110000011
10101010101010101010101010101010101010101010101010

The first sequence (generated by tossing a coin) appears quite patternless. Required to report this sequence of tosses over an expensive communication channel, there would be nothing for it but to say 'The sequence is: "11011..."', spelling it all out digit by digit. By contrast, the second sequence can be reported in a highly compressed way: '"10" is repeated 25 times'. Having a pattern, and hence being non-random – the central KCS suggestion therefore goes – is a matter of this sort of informational compressibility. Conversely, patternlessness is incompressibility. Compression is achieved in our example by providing a short set of instructions for reconstituting the original information. We naturally think of such instructions as encodable as a computer program. So the suggestion becomes this: a binary sequence S is patterned if there is a computer program which specifies it and which is significantly shorter than 'Print: "10110..."', with every digit of S explicitly listed, and hence is significantly shorter than S itself. Conversely, a sequence is *KCS random* if the shortest program for instructing a computer to reproduce the sequence has about the same number of binary bits of information as the original sequence. (Objection: the KCS account has an evident defect – it focuses on a *process* of production, via a short program, yet we have just stressed that our concern is with patternedness and randomness in *products*. Initial reply: the claim is that a KCS patterned string is one that is intrinsically such that it *could* be produced by a significantly shorter program – the modality here is important. The account puts no constraints on how patterned strings *are* produced.)

The suggested definition makes patternlessness vague: just how much shorter must a program be for the sequence S specified by it to count as non-random? Still, the vagueness here is perhaps a desirable feature, rather than a bug: for being patterned comes in degrees, and brevity of algorithm provides a nice comparative measure. When necessary, we can always just choose an arbitrary compression threshold k: e.g. 'if you can't compress the sequence by more than twenty bits, it's random' (it is standard, by the way, to make the compression threshold independent of sequence length; little of what follows hangs on that). Note also that the KCS definition makes all very short sequences 'random' – or better, the definition shouldn't be applied to very short sequences. But again that is a feature rather than a bug. The question whether there is a pattern in a data-sequence doesn't arise for a trivially short run.

Most finite sequences will be KCS random. This follows from simple cardinality considerations. There are 2^n binary sequences of length n, but only $2^0 + 2^1 + 2^2 + \ldots + 2^{(n-k-1)}$, i.e. less than 2^{n-k}, sequences of length less than $n - k$. Hence, even given the most compressed coding of programs into binary strings, there will be less than 2^{n-k} different programs of length less than $n - k$. Hence the proportion of sequences of length n which are specifiable by programs more than k bits shorter must be less than $2^{n-k}/2^n$, i.e. is less than $1/2^k$, a proportion which decreases exponentially with increasing k. Even with the very modest requirement of better-than-twenty-bit compression, less than one sequence in a million of any given length will be non-random.

Given this last point, it is immediate that in our model-of-the-Lorenz-model (§4.2) behaviour is typically KCS random. Label the sheets in that two-sheet structure '0' and '1'. Then if a trajectory starts on the baseline at a point with the binary address $s_1s_2s_3s_4s_5\ldots$, it spends its r-th circuit on the sheet labelled s_r. So the sequence of n visits to the two sheets will be as patterned or patternless as the sequence of the first n digits in the binary address of the starting point. But a binary string of length n is compressible by more than k bits with probability less than $1/2^k$. So the probability is more than $(2^k - 1)/2^k$ that, for an arbitrarily selected initial state, the deterministic operation of the shift map on the string representing the initial state will yield as output a sequence of n leading digits (and hence a sequence of visits to the n sheets) which is KCS random. And though this example involves an artificial model-of-a-chaotic-model, the point carries over to 'real' chaotic models – with suitable binary codings of their behaviour, they generate typically KCS random outputs (see §4).

9.3 Despite the attractive neatness of the basic idea, a number of questions can be raised about the KCS account of randomness. This section comments briefly and inconclusively on perhaps the most interesting of the issues (two other issues are left for comment in the following interlude).

Random sequences, as we have seen, are thick on the ground: nearly all sequences of a given length will count as random. Yet it is quickly shown that there is no effective test for deciding of an arbitrary finite sequence whether it actually *is* KCS random. Suppose there were such a decision procedure, which we can encode into a program Π of length L. Then we could embed this procedure as a subroutine in a slightly longer program Σ which has the following looping structure:

1. Take an initial 'test sequence' consisting of $2L$ 0's.
2. Increment the test sequence (treated as a binary number) by 1, to give S.
3. Run the supposed decision program Π for KCS randomness, to test S.
4. If the tested sequence S passes the test, print it with the comment 'this is KCS random' and halt. Otherwise return to step 2.

Step 1, and the loop-and-increment structure surrounding step 3, require very few, say l, programming steps; so the overall length of the program Σ will be $L + l \ll 2L$. And we know from the cardinality considerations again that some binary sequences of length $2L$ are KCS random. So the program Σ must halt, having – correctly, by hypothesis – correctly deemed some sequence S of length Σ to be KCS random, i.e. not to be the output of any program of length significantly less than $2L$. But that is absurd since, by hypothesis, the program Σ that is printing out the number *is* much less than $2L$ long. (Compare the Berry paradox, concerning 'the least integer not nameable in fewer than nineteen syllables'.) So there can be no such program as Σ. Hence there can be no sub-routine Π for deciding KCS randomness.

Thus, there is no effective test for KCS randomness. It follows from a result of Martin-Löf's that KCS random sequences will tend to *look* as random as a sequence of coin-tosses (Martin-Löf 1966). More carefully, if we suitably match the vagueness in the KCS definition with the stringency of the statistical tests, a KCS random sequence will be statistically random. But the converse result doesn't hold. For example, there is an algorithm for calculating the digits in the binary expansion of π: so the sequence of the first n digits is not KCS random. Yet the sequence of digits of π (continued as far as you like) apparently satisfies statistical tests for random distribution.

Since the narrower property of KCS randomness is not effectively decidable, why not focus instead on the empirically more tractable property of statistical randomness? The following response might be suggested. It is natural to think that S cannot be a truly random sequence when a suitably initiated player can frame a winning strategy for gambling on whether the next digit is a 0 or a 1 as the sequence unfolds. And when a player happens to spot a short algorithm that already compresses an initial segment of S, she may have rational grounds for supposing that the algorithm will continue to apply, and then gambling accordingly. So where there is KCS patternedness, there can be a rational gambling strategy, and hence there is patternedness in the intuitive sense.

However, this line of thought seems to run together exactly what we initially wanted to keep apart (as the chaos theorist must), namely the ideas of randomness of process and randomness of product. When we find that a certain algorithm compresses (an initial segment of) S, it may be rational to conjecture that this is no accident, and that there is some generating process at work which will continue to implement the algorithm as the sequence is extended. Even if our degree of belief in this conjecture is low, it will still be the case that, forced to bet one way or the other on the next digit, it could be rational to bet in accordance with the algorithm, *faute de mieux*. But without *some* non-zero degree of credence in the persistence of a non-random process, we would have no reason to bet that the algorithm will continue to apply (suppose, for example, we had seen the algorithmically compressible sequence actually being generated by chance, e.g. by a series of coin-tosses). Hence our gambling behaviour here, if rational, depends on thoughts, however tentative, about the persistence of pattern in underlying processes. But if that is right, then we can still acknowledge the tie between the availability of a rational gambling strategy and the presence of pattern without being forced to say that there must be patternedness in the product sequence itself as opposed to some regularity in the process generating it.

The question thus remains: why shouldn't we endorse the thought that patternedness in a product should be something there on the surface, which will therefore show up in statistical tests (so it is indeed *true*, as we are inclined to say, that the digits of π are patternless)? In other words, once the distinction has been made, what conclusive reason is there against regimenting the two notions of randomness so that the digits of π count as *product* random, though they can be generated in a simple deterministic way which is *process* non-random? And 'regimentation' is surely the name of the game here: it is not as if we have entirely sharp pre-formal intuitions.

There is a tradition among computer scientists of calling those sequences which pass statistical tests but which are algorithmically generated 'pseudo-random'. The label perhaps misleads. True, such sequences fake it, as far as having a genuinely random source is concerned. But, to press the same question again, why suppose that this in itself means that they consequently fake it on the issue whether there is a pattern left in the data-sequence itself? Without a good answer to that question it seems that we as yet lack a strong reason to prefer the KCS account of finite randomness to, say, some statistical definition in terms of passing a battery of standard tests.

More on KCS randomness

First, a quick observation. This lack of a decision procedure for KCS randomness is directly related to the unsolvability of the 'halting problem'. For if we could (*per impossibile*) effectively decide whether any given program will halt, then we could use a brute force strategy to determine the shortest program that produces a given string *S* of length *n*. Just

1. Alphabetically list in some canonical form all the programs no longer than 'Print [*S*, fully spelt out]'.

2. Throw away the non-halting programs.
3. Run the rest.
4. Chose the shortest which delivers the desired sequence.

But since we can not effectively perform step 2, brute force doesn't work.

The KCS account initially looks elegant and technically smooth; but on closer inspection, at least some of this appearance evaporates. For a start, the suggestion that we define randomness by reference to the relative length of sequence-generating computer programs looks uncomfortably machine dependent. Why shouldn't a sequence which can only be generated by a relatively long program running on machine A have a short generating program running on machine B?

To fix ideas, let's follow Chaitin in considering a dual tape Turing machine M. A 'program' for M is a binary sequence on its input tape. Let $K_M(S)$ – the Kolmogorov M-complexity measure of S – be the length of the shortest binary program fed to M which generates the sequence S on the output tape (if there is no such program, $K_M(S)$ is infinite). Then we may have a pair of machines M and N, and a pair of sequences S and T, such that $K_M(S) \ll K_M(T)$, yet $K_N(T) \ll K_N(S)$. For an extreme case, just choose a pair of dedicated machines D and E whose fixed machine tables ensure they default, on the null input, to generating the sequences D and E respectively (choose machines, that is, whose respective capacities to produce D and E are already built in, rather than being program-specified). Then $K_D(D) = 0 \ll K_D(E)$, yet $K_E(E) = 0 \ll K_E(D)$.

We might hope to minimize this kind of extreme machine dependence by focusing on *universal* Turing machines (which can mimic any machine), and thereby avoid the idiosyncrasies of particular machines. If M is some specific Turing machine which generates S, then in order to produce S from a suitable universal machine U, we can feed U a specification of M (to prime it into duplicating the operation of M) plus the smallest input to M which yields S. This observation yields the result first noted by Kolmogorov, namely that there are machines U such

(K) for all M, there is some constant c_M such that $K_U(S) \leqslant K_M(S) + c_M$,

(the intuitive idea is to set c_M to the length of the specification of M fed to U). Such machines Kolmogorov dubbed 'asymptotically optimal'. In the particular case of the dedicated machine D, the degree of complexity of D is reflected in the size of c_D (which indexes the complexity of D's machine table). In other words, an asymptotically optimal universal machine mimicking D will (unlike D itself) typically need a total input of a complexity matching that of the output D.

Note, however, that (K) only gives an inequality, and this is the best result we can get: for there is nothing to stop there being some sequences S where the total input to U needed to generate S happens to be *much* shorter than $K_M(S) + c_M$ (because U can generate S in some markedly more efficient way than reproducing M's processing). Note also the related point that there are many universal Turing machines, so there need be no unique 'optimal' machine, and that the cases where the machines U1 and U2 are more efficient than some given M can be quite differently distributed. (K), applied to the case where M is another universal machine, tells us that for a pair of optimal machines U1 and U2 there will be some constant c_{U12} such that $|K_{U1}(S) - K_{U2}(S)| \leqslant c_{U12}$. But there is no upper bound on possible values for c_{U12}: there is scope, then, for two asymptotically optimal machines still to differ very markedly in their verdicts on the complexity ranking of a set of binary sequences of given length (and to differ over which sequences have complexity more than *k* below the maximum and so count as non-random).

Moreover, we haven't yet escaped from the possibility of extreme cases of the kind noted before. For if U is asymptotically optimal, then so is each U_D which is constructed by adding to U's machine table further instructions simply dedicated to printing out some *D* in response to the null input. Trivially, each such U_D will give its own favoured sequence zero complexity, and so (according to this machine but not to others) the sequence will be as non-random as can be. Now, we can finesse worries arising from these gerrymandered U_D machines by (perhaps reasonably enough) requiring optimal machines also to be minimal in the sense of not having some part of their machine tables especially dedicated to providing in a 'hard-wired' way what could equally be provided by a suitable input program (a minimal machine, then, is to be a program-interpreter and nothing else). But this only excludes the extreme cases; it is unclear how to motivate any stronger restriction that can be proved not to leave room for a quite significant amount of machine relativity in judgements of KCS randomness even among asymptotically optimal machines.

Here's another technical query about the KCS definition: can it be smoothly extended to the infinite case? (It would certainly seem unsatisfactory to advocate one definition for the finite case and an unrelated one for the infinite case). An initially plausible move might be to say that an infinite sequence is KCS_∞ random if every finite initial segment of the sequence over a certain length is KCS random. More carefully, an infinite binary sequence *S* is KCS_∞ random if there is some threshold *k* and minimum *m* such that, for every $n > m$, the shortest program required to output S_n, the first *n* digits of *S*, is at least $n - k$ long.

Initially plausible, but hopeless. *No* infinite sequence satisfies that condition! Start with an initial segment of length *m*: now track through longer and longer initial segments S_n and record the last u digits of each (where $u >> k$). Eventually there must be repetition, as there are only so many sequences of length u; hence for some S_r the final block of u digits duplicates an earlier block. But then there will be a program for producing S_r which uses a single memory register and runs as follows: print [the sequence up to the repeated block], print [the repeated block] and also write it to memory, print [the sequence between the repeated blocks], retrieve [the repeated block] from memory and print. This, and hence the shortest program for producing S_r, will be at most little more than $r - u$ long, i.e. will be notably shorter than $r - k$. So, for any S, it cannot be that *every* initial segment of the sequence S (over length m) is KCS random with randomness threshold k.

Possible patches have been suggested. The simplest is just to weaken the faulty definition and require for KCS_∞ randomness not that *all* initial segments (over a certain length) are KCS random but only that *infinitely many* are. But the carpet still bulges in another place. For on this proposal, sequences will count as KCS_∞ non-random if (all but a finite number of) initial segments are compressible: but there is no requirement that one and the same algorithm compresses the segments of varying lengths. In other words, there could be an unsurveyable sequence of different algorithms that compress different-length initial segments. In such a case, the 'pattern' in a KCS_∞ non-random sequence will be quite inscrutable and unexploitable (though the sequence may, as in the finite case, pass all statistical tests for randomness) – which surely hardly squares with any intuitive notion of patternedness.

But we will not pursue these technical issues any further here: they may be intriguing, but for our present purposes they rather fortunately turn out not to matter very much.

9.4 In the last section, we wondered whether the KCS account of randomness for finite sequences is really preferable to (say) an account in terms of passing effective statistical tests. In the intervening interlude, we noted that there are question marks too over the machine dependence of verdicts on KCS randomness, and over the issue of how to extend it to the infinite case. It is certainly worth raising these various queries, given the decidedly uncritical stance that seems to be passed on from one writer on chaos to another: the KCS account is far from problem-free.

Still, for all that, we will now shelve the outstanding issues here and follow common practice by continuing to appeal to KCS rather than statistical randomness for our default story about (finite) patternlessness. Why the cavalier attitude? Well, as we will now see, the use made of talk of randomness when discussing chaotic systems is in fact quite insensitive to the details in our choice of definition of randomness for finite (or infinite) sequences. It is claimed, crucially, that the output of chaotic models can be as random – i.e. as patternless – as the typical result of a genuinely chancy process. And the evaluation of such *comparative* claims will not, in fact, require fine-grained attention to the precise notion of randomness that is in play. (We can – so to speak – establish whether two phenomena clearly are, or clearly are not, in the same boat without necessarily needing to settle on a precise mapping of the borders of the vessel.)

The central claim is often put like this: chaotic behaviour can be 'as random as a series of coin tosses'. And this is correct, at least if we interpret it with a little care, as follows. Coin-tossing, or so we conventionally pretend, is an indeterministic process, while a chaotic system is deterministic: so if there is an assimilation to be made, it is at the level of product not process. Imagine, then, a pair of idealized black boxes (Fig. 9.1). In one, there is a coin-tossing device – a homunculus, if you like – which records in real time its sequence of tosses on a tape which slowly emerges from the box. In the other box there is a chaotic deterministic system – for the present, let's suppose it realizes our familiar

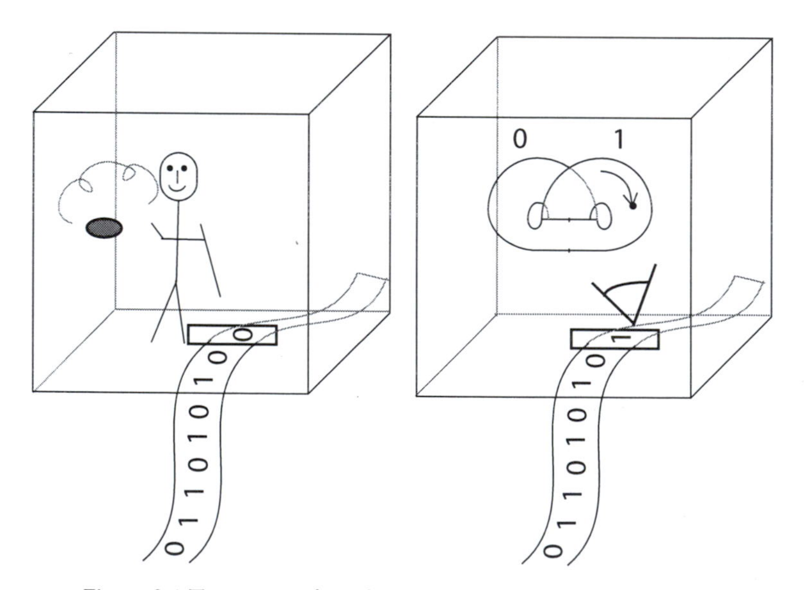

Figure 9.1 Two ways of producing typically random outputs

model-of-the-Lorenz-model, which is set going from an arbitrarily chosen starting point on the baseline. And the tape emerging from this box reports the deterministic sequence of visits by a trajectory to the two sheets in the model. Then, the basic thought is, there will be no way of telling, just by looking at the contents of the tapes as they spool out for any finite time, which box is which, even if we run a number of trials. In particular, there will be an equal likelihood in the two cases of finding patternless outputs of any given length.

Which is indeed true – and it is true irrespective of our preferred account of randomness *qua* patternlessness for finite sequences, whether KCS or otherwise. (Further, the point remains true if we generalize to the infinite case: whatever definition of randomness for infinite sequences we prefer, on a series of trials the infinitary products of our chaotic system will again be random as often as an infinitary sequence of coin-tossings.) But *why* is all this true?

Consider what we mean when we say that our toy chaotic system is set going 'from an arbitrarily chosen starting point'? Suppose there really is a random selection here: perhaps someone tosses (or, so to speak, Nature tosses) a fair coin n times to yield a binary sequence of length n, which is then treated as the initial segment of the binary address of the starting point. This sequence will, in the typical case, be product-random (patternless, on *any* precise definition of the term). Then the resulting trajectory's sequence of visits to the two sheets slowly recapitulates the *same* sequence again; a symbol shift dynamics (§4.2) in effect plays back, bit by bit, the information already encoded in the choice of the starting point. So one of our boxes is recording a series of coin tosses as they happen; the other is in effect playing a time-delayed recording of past coin-tosses (or some other randomizations) that fixed the 'arbitrarily chosen' initial state. It is thus no wonder that we can't tell these black boxes apart just from the patterns on the tape, and that in both cases we get typically random behaviour.

Note that, in a sense, the toy chaotic system is *not* in itself a generator of randomness; for patternlessness is already there in the binary address of the typical starting point. The workings of the dynamics just converts (typically) product-random initial data encoded in a starting point into a (typically) random output data-stream.

Two quick observations. First, we have already said enough to show that Ford's remark that 'chaos is merely a synonym for randomness' must overshoot. Our two black-box systems are as typically product random as each other: but the coin-tossing system is, by hypothesis, not deterministic, and so is certainly not a deterministic chaotic system.

Deterministic chaos is precisely what *distinguishes* some random-product systems from others.

(Is that a cheap shot? Ford, perhaps, meant: chaos is a synonym for what goes on where we have deterministic processing, once the initial state is fixed, plus product randomness. But in fact that is no better: imagine an even simpler third black box, whose initial state is provided by feeding it a long randomly written binary string already printed out: the box just directly spools this string out again. There is trivially deterministic processing and typical patternlessness of product, but this is certainly no chaos machine!)

Second, our imagined chaotic black box contains the artificial 2D-model-of-the-Lorenz-model; however, the set-up is reasonably characteristic in the following sense. In very many 'real' chaotic models there will be a way of partitioning the phase space into two regions – label them '0' and '1' – such that the distribution of the binary sequences recording the sequence of visits to the regions of various possible trajectories is like the distribution of sequences produced by our black box. In particular, with the right partitioning of the phase space, most finite sequences reporting successive trajectory visits are KCS random – and random in any other preferred sense too. (Starting points for trajectories can be classified by the binary strings that record the sequence of partition-visits that they initiate; and on any definition of randomness, most such strings are random.) So, in this way, the chaotic model will indeed produce an output as random as a typical sequence of coin-tosses.

9.5 In our thought experiment, we can not tell the behaviour of the box with the purely chancy insides from the box containing a chaotic system – so how *do* we distinguish, in real-world cases, the patternlessness due to chance and that produced as a chaotic system deterministically unfolds? Answer: we certainly will not be able to tell a purely chancy system from a chaotic one, if all we have to go on is information as exiguous as that available in our thought experiment. But we will normally have significantly more information.

Suppose, for example, our imagined chaotic-system-in-a-box does not just yield up the crudest information about which sheet a trajectory is on during its n-th circuit (in other words, it does not merely print out the leading binary digit of the address where the n-th circuit crosses the base line in the model-of-the-Lorenz-model). Suppose instead it delivers a print-out of (say) the first five digits of each crossing-point address. Then we will have a sequence of successive outputs of the form

$.s_1s_2s_3s_4s_5$, $.s_2s_3s_4s_5s_6$, $.s_3s_4s_5s_6s_7$, $.s_4s_5s_6s_7s_8$, and so on; and it will rapidly become clear beyond reasonable doubt that successive outputs are related by a law $x_{n+1} \approx 2x_n$ mod 1, and hence that the system isn't a merely chancy one. In short, although our chaotic system exhibits product randomness relative to a very coarse-grained description of its behaviour, when we take a somewhat more fine-grained look at the system, we may expose a simple underlying functional relationship between successive states.

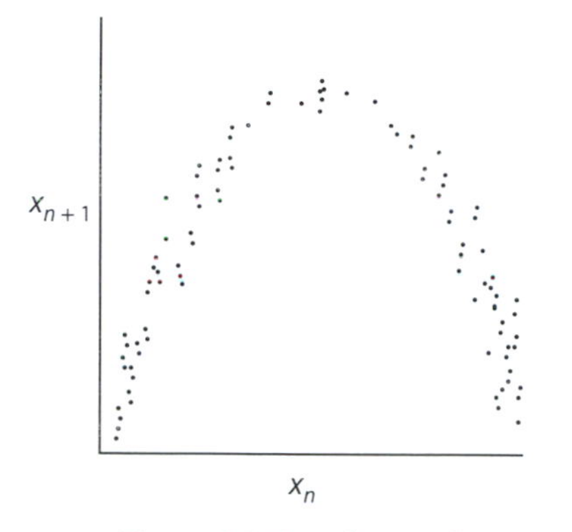

Figure 9.2 Correlation plot

It is similar for more realistic cases. Suppose we measure some quantity x at regular time intervals (or whenever x hits a local maximum, or whenever some other quantity y takes a certain value, or ...). We get a sequence of medium-grained measurements $x_0, x_1, x_2, \ldots, x_n$ – which, let's suppose, we scale to fit into the unit interval. Now it may be that when we impose a suitable coarse-graining on the measurements, we do get a random result. For instance, if we write '1' whenever x_i has a high value, e.g. $x_i > 1/2$, and '0' otherwise, then the corresponding sequence of 1's and 0's is KCS random (or at least, passes statistical tests for randomness). We then naturally ask – is this coarse randomness the result of genuinely chancy behaviour (the effect, say, of indeterministic noise)? Well, try plotting a correlation graph for successive x_i; that is to say, make a 2D plot of the distribution of the points with coordinates $\langle x_n, x_{n+1} \rangle$, as in Figure 9.2. One possibility is that these points scatter all over the place, indicating that the x_i are uncorrelated and hence perhaps that we are dealing with a chancy system (although it *may* just show that we are measuring the wrong values at the wrong times). But another possibility is that the points on the correlation graph approximately sit on a nice curve, as in the Figure, suggesting that there's a simple mapping relation $x_n \Rightarrow x_{n+1}$.

To take a now familiar example, it might perhaps be that the correlation approximates the logistic map $x_{n+1} = 4x_n(1 - x_n)$. This map is deterministic, but almost any starting point leads to coarse-grained randomness. In other words, from typical starting points, the sequence of visits to the partitions [0, 1/2], (1/2, 1] will be random. (To see this, consider again the transformation $x_i \Leftrightarrow 1/2.(1 - \cos a_i\pi)$ discussed in §4.2, which relates the logistic map to a symbol shift dynamics. Since $x_i \in [0, 1/2]$ iff $a_i \in [0, 1/2]$, the sequence of visits to the partitions by the x_i is exactly matched by the sequence of visits of the a_i. But this pattern will just be the pattern of the leading digits thrown up by a symbol shift dynamics working on the initial a_0. And this pattern, for arbitrarily selected x_0 and derived a_0, will typically be as random as a sequence of coin tosses.)

In sum, faced with coarse-grained randomness in some empirical product, we may yet be able to find a level of finer-grained order, showing that we are not dealing with a merely chancy, indeterministic regime. True, if there is very complex behaviour in an empirical case, it may be difficult to locate the right quantities to measure and hence difficult to decide whether the surface complexity and apparent randomness in fact arises from some underlying deterministic chaos. But still, there seems no ultimate difficulty of principle in distinguishing the presence of chaos from mere noise.

9.6 Note, though, that while empirical chaos can be distinguished from *mere* noise, it is (almost certainly) never entirely noiseless. Relatedly, there is no pure deterministic randomness in the world.

Take an experimental set-up where we can empirically produce what seems to be long-term chaotic behaviour – say in the BZ reaction (Chapter 8). We saw that we can construct from the empirical data-sequence a representation of the dynamics with the right kind of stretching and folding of phase space trajectories for chaos. Now, where we have 'stretching', we have the inflation of small initial differences: so the later digits in the address of a point representing an initial state become more significant over time. And 'folding' in effect throws away information about leading digits. That's why stretching and folding gives rise to a symbol-shift dynamics; cf. §4.2 again. Since (the initial segments of) most strings of symbols representing initial states are random, a symbol-shift dynamics will in turn yield typically random behaviour. So our *model* of the BZ reaction implies that we will find deterministic randomness.

But the BZ reaction cannot truly exhibit pure deterministic randomness of the kind we found e.g. in our toy chaos box. For we cannot really interpret some very long stream of binary data gathered sequentially at (say) t_1, t_2, t_3, ... t_{100} as deterministically unfolding the initial segment of the expansion of some random real indexing the precise initial starting point for the system at t_0. For it will simply not make physical sense to suppose that the initial quantities concerned here – i.e. proportions of chemicals – have the corresponding hundred-bit precision (see §3.2). Physically, as we have emphasized before, what is going on is that there is a dynamic evolution of a stretching-and-folding kind which *if* executed with perfect precision *would* produce true chaos – but in fact it is a noise-perturbed and imprecise evolution, and randomness in the data-stream will have a mixed origin. Likewise for other candidate cases of real-world chaotic randomness.

Still, that is certainly no reason to give up using standard models built in the real number system with all its excess precision. To return to the 'robustness' point of §7.5: faced with the coarse-grainedness of the quantities whose time evolution we want to model, our best theoretical move may well be to represent coarse-grained physical quantities by real numbers; and then we take seriously the results which are robust in not depending too sensitively on our precisifications (and which are also stable with respect to small amounts of added random noise). Among such robust features, preserved across a family of acceptably successful models, may be having a stretch-and-fold dynamics which inter alia can generate patternless products. When this is so, it will be apt to talk of the empirical randomness we find as being in part due to deterministic chaos. Even if there isn't pure deterministic randomness in the worldly system being modelled, at least there is not-purely-chancy randomness.

A corollary: suppose talk of worldly almost-deterministic randomness *is* to be grounded in the rough but robust applicability of certain idealizing models with deterministic randomness. Then the prime notion we need to analyse is the notion of product randomness applicable to the mathematical models – and *that* can be an infinitary notion, as the models have infinite outputs. So one putative key merit of the KCS account, namely that it directly applies to the finite case, is after all no special recommendation.

The modelling medium

We saw that the typical randomness in *output* of e.g. our toy chaos box comes for free with the typical *input* numbers which might specify a starting state. A worry is sometimes voiced: is the randomness to be found in chaotic models thus merely an artefact of the modelling medium, without worldly significance? The discussion of the last two sections should allay any worry of this kind. But it is worth adding a brief comment.

There is a danger of distorting the delicate relationship between properties of chaotic systems and properties of their representations in the reals. It is convenient for most purposes to talk as if a chaotic model lives in some space that is R^n (better: '... in a space which is piecewise R^n' – this qualification allows for models that, say, live on circles or spheres rather than lines or planes, but henceforth we will ignore it). But, strictly speaking, our models should really be regarded as living in more abstract phase spaces, sets of points which form *n*-dimensional differentiable manifolds – manifolds which can be *represented* by R^n but are not to be *identified* with R^n, since there will be many different representations of a given manifold in R^n (shift the origin, spin the axes, double the real-number address of every point ...). This differentiable structure provides the true modelling medium, the setting in which we can define dynamical flows which continuously map points to other points and so induce phase space trajectories; and we can characterize sensitive dependence on initial conditions, aperiodicity and confinement (the usual marks of chaos) just in terms of the basic metrical structure. Further, we can (for example) partition the phase space and say such things as: the sequence of visits of the partitions by a typical trajectory is patternless. In the right sort of case, we may be able to give a specification of any point in some region by giving the infinite sequence of visits to the partitions of a trajectory starting from that point. Then, relative to such a specification of initial points, the dynamics will then induce a symbol shift map of the kind already discussed. And none of this depends on exploiting some representation of the points in the manifold by *n*-tuples of reals.

So it isn't at all clear that we should say (as it sometimes is said) that the facts about chaotic randomness *depend* on e.g. facts about how most reals are random. It would, perhaps, be better to put it this way: it is general facts about the incompressibility of most binary sequences that *both* allow us to say that chaotic systems can generate product-randomness *and* imply that real numbers can be said to be typically random. Though it is, to be sure, also unclear what the content of any such dependency claims is, given we are dealing with non-contingent facts, which in a sense depend on nothing.

10

Defining chaos

10.1 And what, finally, *is* chaos? So far, we have taken a few paradigm cases of mathematical models with complex behaviour (e.g. the Lorenz system, the logistic map), noted their intricate features, and then asked some key questions about the role that such infinitely structured models can have in representing a messy world and explaining natural phenomena. Answering those key questions – the central concern of this book – does not at all depend on having a sharp official criterion for separating the strictly chaotic from the non-chaotic cases.

A quick glance at the research literature (in journals like *Physica D*) shows that working applied theorists also proceed without any agreed precise definition of what counts as 'chaos'. Rather, the term is typically used quite loosely, to advertise the presence of some interesting cluster of the phenomena that we have illustrated – e.g. exponential error explosion, the existence of a fractal attractor, the equivalence to a 'symbol shift' dynamics with product-random output, and so forth. Still, there is some interest in reviewing various options for giving a tidy definition (and some philosophical interest in reflecting on the nature of this definitional enterprise).

In this chapter, then, we consider possible definitions of chaos for (the dynamics in) a mathematical model. We have already offered a rough indication of how, given such a definition, we can extend the notion to apply to some real-world dynamical behaviour. To draw the threads together again, suppose a real-world phenomenon has an acceptable mathematical model (i.e. acceptable by the canons of allowable idealization, simplification, etc., that usually guide model selection), where this model has whatever feature or features makes it count as chaotic according to our preferred definition. Or rather better, suppose there is a *family* of empirically acceptable models, which may differ in the exact parameter settings in their defining equations, in the level of allowance for random noise interference, etc.; and suppose that the chaos-making features are robustly present across this family of acceptable models. Then we may deem the real-world phenomenon to be chaotic in an empirical sense. Which – to repeat the key point stressed before – isn't to say that the chaos-making features of *models* can themselves be found directly instantiated in the *world*. For example: assume that we take chaos in a model to involve at least sensitive dependence on initial conditions defined as in (SDIC), §1.5. And suppose that any adequately competent model of e.g. Rayleigh-Bénard flow (for certain settings of the experimental parameters) has sensitive dependence. Then, assuming any other necessary conditions are also satisfied, we may deem the flow, for those settings, to be chaotic in the empirical sense. But this does not require that the physical quantities, *per impossibile*, take the infinitely precise values which would be required if the ϵ–δ definition of (SDIC) were strictly to apply to *them*.

10.2 Recall the link between difference maps and models with continuous trajectories (§6.1): if we transect a trajectory bundle with a suitable surface of section P, the dynamics induces a discrete map on P – namely, the recurrence map which takes a point x_n on P to the point $x_{n+1} = f(x_n)$ where the trajectory through x_n next hits P. The link suggests a possible approach to our definitional task: first characterize chaos for discrete maps, and then suitably extend the definition to the continuous case (so that where a return map induced on some surface P is chaotic, the original continuous dynamics counts as chaotic too). This approach has some attractions. Discrete maps provide a simpler setting in which to first encounter some of the complexities in defining chaos. And there is a rich but relatively accessible literature on the discrete case. So this is the approach that we will explore.

If the map's defining function f is allowed to be very 'wild', then of course its iterates can wander in a correspondingly wild way. Our concern, however, is to pin down the sort of complex behaviour that can arise even when f is thoroughly well-behaved (as in e.g. the logistic map). So, as is standard, we will concentrate on cases where f is at least a continuous function in its domain. In this section we will explain and motivate the definition of chaos given by Robert L. Devaney in his widely influential *An Introduction to Chaotic Dynamical Systems* (1989): in §10.4 we will meet three alternatives.

A reminder: the logistic map defined by $g(x) = 4x(1 - x)$ has uncountably many aperiodic orbits confined to the unit interval $[0, 1]$, and these orbits typically wander unceasingly all over the unit interval (see §6.2). That is to say, the attractor for the dynamics is no less than the whole interval. This simple observation is enough to sabotage any idea that having a 'strange' attractor is necessary for chaos: there is paradigm chaos in the logistic case, even when the attractor is a simple interval. And incidentally, having a strange attractor isn't a sufficient condition for chaos either – there can be dynamics defined over the unit interval where the attractor is fractal but where there is no chaos (call this Result A: labelled results get proof-sketches in the next interlude). In sum, chaos is a feature of the dynamics itself which isn't dictated by the nature of the set over which the dynamics is played out.

Let us return, then, to our initial informal characterization of chaotic dynamics in terms of a cluster of features – sensitive dependence, a mix of periodic and aperiodic behaviour, etc. (§§ 1.5, 4.3, 6.2) – and consider ways of sharpening these ideas.

We need briskly to (re)establish some notation and terminology. Suppose f is a function defined on S. We write $f^n(x)$ for the result of re-applying the function n times, i.e. $f(f(f(\ldots f(x))))$. If K is a subset of S, then we write $f(K)$ for the set of points $f(x)$ where $x \in K$: i.e. f maps the set K to the set $f(K)$. If $f(K) = K$ then we say K is invariant under f. And an attractor for f is a closed invariant set A such that any orbit starting close enough to A approaches A arbitrarily closely in the limit.

We might, as a first shot, think of characterizing a map f as chaotic if it has an attractor on which the dynamics is chaotic, whatever that turns out to mean exactly. But on second thoughts, why not be a little more liberal? If f has an invariant set K on which the motion is chaotic, then why not say that f exhibits chaos, even if K isn't an attractor and doesn't (so to speak) dominate the local dynamics? For the present, we will take this more liberal line. So what we now need to define is the notion of a dynamics being chaotic over some (invariant) set K.

First, and crucially, we require some kind of sensitive dependence. The analogue for discrete maps of the original definition (SDIC) for continuous systems is as follows: a map f is (weakly) sensitively dependent on K if there exists an $\epsilon > 0$ such that, for every x in K,

$$(\mathrm{SDIC}_m) \quad (\forall \delta > 0)(\exists y \in K)(\exists n)(|x - y| < \delta \text{ and } |f^n(x) - f^n(y)| > \epsilon).$$

That is to say, there is (weak) sensitive dependence on the set K if, in any neighbourhood of a given point, however small, there is some other point y belonging to K such that the orbits starting from x and y eventually peel apart by at least ϵ.

As we noted for the original (SDIC), this *is* in itself quite a weak condition – in particular because it says nothing about how fast the orbits in a neighbourhood peel apart. Stronger versions of (SDIC_m) will explicitly require e.g. that orbits that start close together peel apart *exponentially fast*. One such version might be

(L_m) if $x \approx y$, then $|f^n(x) - f^n(y)| \approx |x - y|e^{\lambda n}$ where $\lambda > 0$.

But as also noted before, even strong sensitive dependence properties certainly do not suffice for chaos (cf. §1.5 for the continuous case). Take the dynamics defined over $[0, \infty)$ by the function $f(x) = 2x$. This is sensitively dependent in any plausible sense – the gap between orbits starting at points x_0 and y_0 doubles at every step – but the dynamical behaviour has no interesting complexity at all.

Of course, iterates of $f(x) = 2x$ boringly accelerate off to infinity. By contrast, the signature of chaos, we suggested, is the combination of sensitive dependence with the further structure that comes when continuous trajectories or discrete orbits are 'folded back' on themselves. In familiar cases – the Lorenz model, the logistic map – such folding back is enforced by the fundamental fact that trajectories or orbits are (eventually) trapped in some finite region of phase space. So should we lay down, as a condition for chaotic motion on the set K, that the set be bounded?

Interestingly, this condition is often *not* added (and it isn't added by Devaney). But why so? Here is a possible motivating line of thought. Suppose we have a slight variant on logistic map g – call it g_o – whose iterates are confined to the *open* interval $(0, 1)$ and whose typical orbits wander chaotically over that interval, approaching 0 and 1 arbitrarily closely. Imagine now a simple change of coordinates, e.g. $x \Rightarrow (1 - x)/x$, that takes points in the interval $(0, 1)$ to points in $(0, \infty)$. In the new co-ordinates, typical orbits of g_o wander chaotically *and unboundedly* over the half-line, since values arbitrarily near 0 in the old coordinates get converted to arbitrarily large values in the new coordinates. Which seems to give us chaos without finite confinement. (For the construction of a suitable g_o, see Result B.)

The moral is that what matters for chaos (whether in the continuous or the discrete case) may not be the confinement of trajectories or orbits so much as the folding back which is typically implied by confinement, though does not require it. So let's not build confinement into our definition, but rather focus on that key implication.

'Folding' is more than a mere metaphor, but it is hardly a sharp idea. In §10.4 we will see one rather natural way of giving the idea precision in the one-dimensional case. But for the present, consider what folding, intuitively understood, gives rise to. So take a bunch of orbits, starting in some small neighbourhood $N \subset K$, which both spreads apart to give sensitive dependence and eventually (after, say, k steps) is 'folded back' to cover the originating neighbourhood. If f is continuous, so is f^k: but if, to put it crudely, the effect of f^k is to smear N continuously across itself, then we will expect some points to be mapped exactly to themselves. In other words, there will be some periodic points in N. And it is easy to check that in e.g. the logistic case for $\mu = 4$, there are indeed periodic points in *every* neighbourhood on the attractor, i.e. *the periodic points are dense* on the attractor (Result C, already noted in passing in §6.2). This seems typical of intuitively chaotic cases. So we might try adding, as a condition for chaotic dynamics on K, the presence of a dense sprinkling of periodic points across K. (This immediately rules out the function $f(x) = 2x$ from counting as chaotic on $[0, \infty)$.)

Certainly, the co-presence of sensitive dependence and a dense covering of periodic points imposes severe conditions on a dynamics. But is it enough for chaos? Consider next the following artificial case, a discrete dynamics defined this time over the unit square, which maps a point with coordinates (x, y) to the point $(g(x), y)$, where g is the logistic map again. Given what we know about g, it is immediate that the attractor for this dynamics is the whole unit square; there is sensitive dependence; and there is a dense covering of periodic points. But it seems plausible to say that the dynamics here is not truly chaotic motion over the whole square – rather we have simply bolted together a chaotic dynamics in one dimension with an entirely static 'dynamics' in the other. Plausibly, where some motion is to count as chaotic right over a set K, we want a thorough mixing up of orbits – that is to say, a bunch of orbits starting near any given point u should set off wandering all over the place, with at least some of them getting arbitrarily close to any other given point v. That doesn't happen in our 'square' example: any orbit starting near (x_0, y_0) always keeps its original y value.

Let's put the 'mixing up' requirement more carefully. Suppose f is a function that maps K to K. Then for truly chaotic behaviour right across K, we might plausibly want the following. Pick a neighbourhood, or more generally an open set, U round the point u; and pick an open set V round any other point v: then, however small U and V are, *some* orbit starting in U eventually visits V. If this holds, we say f is *topologically transitive* on K. (We needn't worry about the significance of the label. Again, it is easy to see e.g. that the logistic map for $\mu = 4$ has topological transitivity on its attractor – that's Result D.)

We can now state the Devaney definition of chaos. A continuous map f defined on S is chaotic$_d$ if f has an invariant set $K \subseteq S$, such that

(1) f is (weakly) sensitively dependent on K,
(2) periodic points are dense in K,
(3) f is topologically transitive on K.

This definition has been widely adopted by later writers: but is it acceptable? It turns out – though this has only relatively recently been noticed (Result E) – that conditions (2) and (3) imply (1). But, in itself, such redundancy is perhaps only an aesthetic flaw. The interesting question is whether the remaining conditions should be accepted as necessary and sufficient for chaos. Is chaos$_d$ chaos?

Some results on maps

Partly to help fix ideas, and partly for fun, this interlude gives quick explanations of the various mathematical Results just noted.

First, to establish Results C and D: the logistic map $g(x) = 4x(1 - x)$ has dense periodic points and is topologically transitive. Recall the intimate relation between the behaviour of this map and a symbol-shift (or rather, shift-and-flip) dynamics defined over infinite binary expansions – see §6.2 again for the details. It is enough, then, to establish the analogous Results C* and D* for this equivalent dynamics. Suppose, then, that a, a' are in $I = [0, 1]$, with $a < a'$. Their binary expansions must eventually diverge, so that we must be able to interpolate a pair of binary numbers c^-, c^+, such that

$$a < c^- = .c_1c_2c_3\ldots c_n0\overline{0} < c^+ = .c_1c_2c_3\ldots c_n1\overline{0} < a'$$

where overlining indicates infinite repetition. Now consider the point

$$c = .c_1c_2c_3\ldots c_n\overline{0c_1c_2c_3\ldots c_n0}$$

By construction, $a < c^- < c < c^+ < a'$, and c gives rise to a repeating orbit of the shift-and-flip dynamics. Hence, between any pair of points in I, there is indeed a periodic point for the latter dynamics, which is Result C*.

Next, suppose U is an open set containing u; then within U there must be some open neighbourhood (a, a') straddling u. Choose a c^- and c^+ as before, and then construct

$$v^* = .c_1c_2c_3\ldots c_n0v_1v_2v_3\ldots v_n\ldots$$

where $v_1v_2v_3\ldots v_n\ldots$ is the binary address of some arbitrary point v. We have $a < v^* < a'$, so $v^* \in U$; and the shift-and-flip dynamics eventually takes the orbit beginning at v^* to the point v (and so, a fortiori, takes the orbit into any arbitrarily small neighbourhood containing v). Which is enough to show the topological transitivity of the shift-and-flip map, which is Result D*.

Similar arguments work whenever a dynamics is equivalent to some species of symbol shift dynamics (maybe involving more than two symbols,

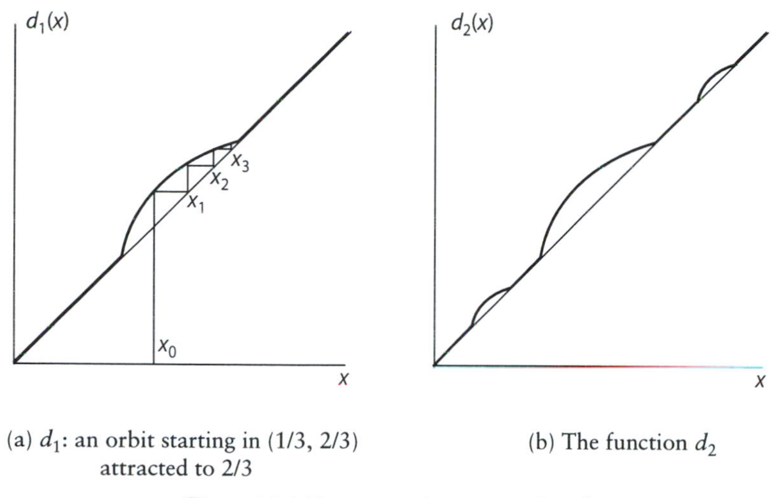

(a) d_1: an orbit starting in (1/3, 2/3) attracted to 2/3

(b) The function d_2

Figure 10.1 First stages in constructing d_∞

maybe involving some restrictions on which symbols can follow which, etc.). The equivalence is particularly easy to see in the case of the logistic map; but such equivalences can be found much more widely. For we can often partition an invariant set so that the dynamics of an orbit is fully described by giving the sequence of visits to the partitions, in which case the dynamics can be thought of as operating like a shift map over a sequence of symbols labelling the partitions (cf. §4.2 and the final interlude).

Next, we establish Result A, that there can be non-chaotic dynamics with a 'strange' attractor. Consider the sequence of functions d_0, d_1, d_2, ..., d_∞ defined over the unit interval (I am following here a recipe suggested by Peter Dixon). d_0 is the identity function. The graph of d_1 is illustrated in Figure 10.1a – we cut out the middle third of the diagonal graph for d_0, and replace it by a (monotonic increasing) curve above the diagonal. Graphical analysis of the behaviour of iterates of d_1 – see Figure 6.8 and commentary for the technique – shows that orbits starting in the interval (1/3, 2/3) are attracted to the point 2/3; all other points, by definition, are mapped to themselves. To put it another way: the set of fixed points for d_1 is the 'pre-Cantor' set C_1 (see §2.3), and one of these fixed points is the attracting point for all orbits starting outside C_1. To construct the graph of d_2, now cut out the middle thirds of the straight parts of the graph of d_1, replacing these by curves above the diagonal (Figure 10.1b). The fixed points of d_2 comprise exactly C_2; and orbits starting at outside C_2 are attracted to points in C_2. Now keep on going in the obvious way, to construct d_∞. The set of fixed points for d_∞ is the Cantor set C_∞; and all orbits starting outside that set are attracted to points within it. In

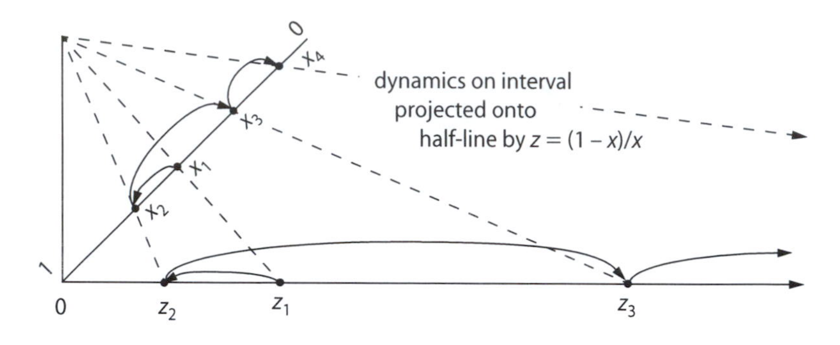

Figure 10.2 Projecting dynamics from (0, 1) to (0, ∞)

fact, C_∞ is readily seen to be the attractor for the dynamics. So here we have a fractal attractor: but the dynamics on C_∞ is certainly not chaotic – every point in C_∞ is a *fixed* point!

Now for the possibility of unbounded chaos, Result B. We can 'project' a dynamics defined on the open interval (0, 1) onto the open half-line $(0, \infty)$, using a simple transformation such as $\tau{:}\, z = (1 - x)/x$, as in Figure 10.2. If f gives rise to a chaotically complex dynamics on the interval, then its projection $f^*(z) = \tau(f(\tau^{-1}(z)))$ surely gives rise to an equally chaotic dynamics: a simple change of coordinates shouldn't matter for chaos. Thus, for example, suppose the original map f is the logistic map defined on (0, 1) with μ just above $c_\infty \approx 3.570$, where the motion first becomes chaotic and there is a fractal attractor (see §§6.3, 6.4): then the projection f^* gives us chaotic motion again, governed by a new fractal attractor.

So far, so unexciting (for in the last case, the new attractor is still a finitely bounded set). But consider next a case based on the logistic map g for $\mu = 4$, where the dynamics is chaotic and the attractor is the whole interval. We get into trouble if we try to project the full dynamics of g itself (some points initiate orbits which get mapped to $0.5 \Rightarrow 1 \Rightarrow 0 \Rightarrow 0 \Rightarrow \ldots$; but the projection of such orbits by τ is undefined since $\tau(1)$ is undefined). Still, consider instead the set J which is the unit interval minus its end points *and minus the denumerable number of points which initiate orbits of g which eventually visit the point 0*. And let g_o be the restriction of g to J. Then J is invariant under g_o, and the dynamics of g_o on J is surely just as complex as the original dynamics of g on the closed unit interval (we have simply lost a handful of eventually periodic orbits which end up at 0). So g_o intuitively has a chaotic dynamics – and indeed is chaotic$_d$ by Devaney's definition too. Now project *this* dynamics into $(0, \infty)$, and we get a dynamics which is as chaotic as before but where orbits can wander without bounds – since values of g_o get arbitrarily close to

0, i.e. can visit points $x < 1/(N+1)$ for any N, values of g_o^* can become arbitrarily large, i.e. can visit points $z = \tau(x) > N$ for any N. Hence there's chaos (and indeed chaos$_d$) without confinement.

Finally, Result E: the second and third conditions of Devaney's definition entail sensitive dependence (first proved in Banks et al. 1992). Suppose then that (2) f's periodic points are dense in K and (3) f is topologically transitive on K. What we will show is that, near any point x, there must be a pair of points p and q whose n-th iterates $f^n(p)$ and $f^n(q)$ (for suitable n) are 'far' from each other. Then $f^n(x)$ must be 'far' from one or other of $f^n(p)$ and $f^n(q)$, and hence one or other of the pairs x-p, x-q, verifies the sensitive dependence at x.

The brisk details which follow can certainly be skipped: they are given mainly because this isn't yet a textbook proof.

By (2), f has an infinite number of distinct periodic orbits. Take a couple of such orbits, o_1 and o_2, and suppose that the minimum distance between a point in o_1 and a point in o_2 is 8ϵ. Then any point x in K must either be at least 4ϵ from every point in o_1, or else at least 4ϵ from every point in o_2. Choose the further orbit o from x, and let r be a point on that orbit.

Now to find our 'p' and 'q'. Choose a δ, arbitrarily small (without loss of generality, we'll assume it is smaller than ϵ). First, since periodic points are dense, then there must be periodic points within δ of x: let p be one, with period π. To fix q is just a little more complex. As we'll check below, there exists a (non-empty) open set R_π of points which are (a) within ϵ of r and (b) whose next π iterates also all stay within ϵ of iterates of r (i.e. stay near the orbit o). By transitivity, there must be within δ of x some point whose orbit eventually, after (say) k iterates, visits R_π. Let this be q. Now set n to be the first exact multiple of π greater than k. Since n is an exact multiple of π, $f^n(p) = p$, a point within δ – and hence within ϵ – of x. Since iterates of orbits that hit R_π stay within ϵ of the orbit o (which is always at least 4ϵ from x) for at least π iterations, that means that $f^n(q)$ is at least 3ϵ from x, i.e. at least 2ϵ from $f^n(p)$. So, wherever $f^n(x)$ ends up, *that* must be at least ϵ from either $f^n(p)$ or $f^n(q)$. Which verifies (SDIC$_m$) by showing that, however small δ is,

$$(\exists y \in K)(\exists n)(|x - y| < \delta \text{ and } |f^n(x) - f^n(y)| > \epsilon).$$

Finally, to check that the requisite R_π exists, let R_0 be the open set of points in K within ϵ of r. Let R_1 be the intersection of R_0 with the open set of points whose first iterates are within ϵ of $f(r)$ – whence R_1 is open and $R_1 \subseteq R_0$, and iterates of points in R_1 stay within ϵ of orbit o for at least one step. Let R_2 be the (open) intersection of R_1 with the set of points whose second iterates are within ϵ of $f^2(r)$; and keep on going similarly to construct R_π. R_π is open, it is non-empty as it contains r, and its points iterate within ϵ of o for at least π steps. QED

10.3 Before getting entangled in yet more mathematical details – and we must, given we are discussing how to construct a mathematical definition! – let's pause to consider what sort of question it is, whether Devaney's definition is acceptable, whether chaos is chaos$_d$.

Definitions in mathematics get shaped by a number of pressures. We may start with a cluster of informal basic results which we want our formal definitions broadly to sustain (some of the intuitive results may be non-negotiable; others may be up for possible revision). There is then, on the one hand, the desire for increasing generality, inclusiveness, abstractness. But on the other hand, we also want the defined concepts to feature in powerful theorems (a suggestive analogy: physical natural kinds are those which feature in a rich network of causal law, mathematical 'natural kinds' are those that feature in a rich network of substantial theorems). Connectedly, we want there to be interesting relations to (refinements of) other, well-entrenched, mathematical concepts.

In his classic treatment of this theme, *Proofs and Refutations* (1976), Imre Lakatos explores, in particular, the way that concepts are shaped and reshaped to secure the validity of key proofs. His central example concerns the notion of a polyhedron and its development from a naïve to a 'proof-generated' concept which is needed if the Euler conjecture about the number of vertices, edges and faces ($V - E + F = 2$) is to be provable as an exceptionless theorem. But this example, where the enquiry is dominated by the aim of assessing a single major conjecture, perhaps isn't the best model for our purposes. So consider an example from another domain which Lakatos also (very briefly) touches on.

In measure theory, we develop conceptions of 'measure' that enable us to talk about the measure not just of simple regions but, more generally, of 'Borel sets' – very roughly, sets formed by repeatedly taking unions or intersections (starting from interval-like regions and their complements); and then, pressing on, we may want to talk about the measure not just of Borel sets but of arbitrary subsets of R^n. In parallel to these increasingly general conceptions of measure, we can develop e.g. increasingly general notions of integration, allowing us to make sense of integrating functions over sets which aren't simple regions. These various notions of measure sustain a spectrum of general theorems of differing strength – though in each case, a cluster of intuitive basic results (e.g. 'the empty set has zero measure', 'the measure of [0, 2] is the sum of the measures of [0, 1] and [1, 2]', etc.) stay fixed.

The desire for generality and the desire for a rich network of theorems evidently push in different directions, for the more general and all-embracing a concept, the fewer the interesting truths about all its instances. And hence there may be a number of acceptable ways of trading off the virtue of generality against the virtue of theorem-generation, and a number of concepts encapsulating these different trade-offs. Thus, there turn out to be useful notions of Borel measure, Lebesgue measure, 'outer measure', and so on (likewise, there is the Riemann integral, the Lebesgue integral, and so on). More obviously than in Lakatos's polyhedron example, there is certainly no one 'right' or 'best' concept of measure (or integral).

Another quick example, closer to our current concerns. We had occasion in Chapter 2 to introduce, very informally, a number of notions of dimension (over and above the 'school geometry' notion). We introduced, for instance, the idea of box-counting dimension, and we also mentioned – though we didn't explain – Hausdorff-Besicovich dimension (§§2.3, 2.4). Again, there is no one 'right' concept of dimension apt for the description of monstrous fractal sets. Rather there are trade-offs. We saw (in the interlude 'Fractal dimension') that are cases where box-counting gives us 'unintuitive' results, assigning a fractional dimension to some very 'tame' sets. On the other hand, it is usually much easier to establish the box-counting dimension of a set (which, in any case, very often equals the Hausdorff-Besicovich dimension); and we can prove some nice theorems about box-counting. So in fact mathematicians continue to make use of both notions.

We might reasonably expect the notion of chaos to be similarly susceptible to a number of alternative regimentations, corresponding to differently satisfying trade-offs between generality, theorem-generation, connectedness with other notions, and the preservation of various informal judgements of chaotic 'monstrosity'. And this expectation may be bolstered by the thought that (as with 'measure' and 'dimension' but perhaps unlike 'polyhedron') we do not start off with a single canonical set of conjectures that we want our developed concept of chaos to help establish as theorems.

In short, there need be no one 'right' or 'best' concept of chaos.

10.4 The question whether chaos is $chaos_d$ is thus better put: does Devaney's definition hit on *one* reasonable regimentation of our pre-formal ideas?

As we noted, in its original version, the definition has a redundant clause. So we can say, more simply, that f is $chaotic_d$ just when f has an invariant set K such that periodic points are dense in K, and f is topologically transitive on K. That still looks brutely conjunctive:

however, we can improve matters. For Devaney's definition is in fact elementarily equivalent to the following:

> f is chaotic$_d$ just if there is an invariant set K such that every pair of (non-empty) open subsets of K shares a periodic orbit.

That is to say, take subsets U, V; then there is a point in U which initiates a periodic orbit which eventually visits V. (Call this equivalence claim Result F; it is proved in the next and final interlude.)

And now, in this final version, Devaney's account might well strike us as rather surprising as a basic definition of chaos. After all, at the outset (e.g. in §1.4) what we emphasized about paradigms like the Lorenz model was the presence of *aperiodic* behaviour, with typical trajectories peeling apart and – although confined – never quite repeating. So, we might ask, why now the emphasis on the presence of the *periodic* orbits? (For it is not as if we can drop the qualification 'periodic' from the last definition: it is easy to find maps which are not chaotic, but where every pair of open subsets share an orbit – that's Result G.)

To press the last point, consider two more variations on the logistic map g defined over the unit interval $I = [0, 1]$. Let $P \subset I$ be the set of periodic points for g, and let $A \subset I$ be the set of aperiodic points. Let g_A be the restriction of g to A and g_P be the restriction of g to P. Then, firstly, $g_A(A) = A$, and motion on this invariant set still seems intuitively chaotic (there is a complex tangle of aperiodic orbits, sensitive dependence, and confinement ...). Yet it isn't chaotic$_d$ – far from the periodic orbits being dense on A, g_A has no periodic orbits at all. On the other hand, g_P *is* chaotic$_d$ on its invariant set P (any pair of open sets in P will share a periodic orbit). But do we really want to say that g_P is truly chaotic, despite having *no* aperiodic orbits? Well, perhaps this is a 'don't-care case', where intuitions give no firm verdict and we can let best theory tutor our pretheoretic judgements. But some will feel that this too isn't an entirely happy result.

We can finesse these worries if we take our concern to be defining chaos for mapping functions which are not only 'nice' in the sense of being at least continuous in their domains but also have domains which are 'nice' as well (e.g. are defined everywhere in some region of R^n) – so functions like g_A and g_P with gappy domains are not strictly candidates for true chaos. Yet even if we take this 'monster-barring' gambit (to borrow a phrase from Lakatos), we are still left with the Devaney definition's somewhat surprising emphasis on the periodic orbits. We may well begin to suspect – even if it is unclear just what kind of judgement this is – that being chaotic$_d$ is a typical *consequence* of chaos, rather than the basic mark of chaos. So what are the definitional alternatives?

In the rest of this section, we will briefly indicate three possible lines of enquiry – labelled (1) to (3) below. And to keep the burgeoning complexities under some control, we will now concentrate on the simplest case where f is a one-dimensional map of an interval I into itself (the possibility of simple rescaling means that we can think of this, without loss of generality, as the unit interval).

(1) In §10.2, we noted a typical implication of the 'folding' of orbits, namely the presence of a dense covering of periodic orbits on an invariant set – and we built this implication into the definition of chaos$_d$. But rather than focus on one of its consequences, why not try to capture more directly the basic idea of stretching and folding that seems so characteristic of chaotic dynamics?

If some function h 'folds' orbits, then we will expect there to be non-overlapping sets M_0, M_1 which get mapped into the *same* set N. As for 'stretching', if there is sufficient spreading apart of orbits, then N may in fact be large enough to contain both M_0 and M_1 as subsets. In the case where h is a map of the interval I into itself, and M_0, M_1 (and so N) are themselves open intervals, this effect is as illustrated in Figure 10.3 (cf. other 'horseshoe' diagrams, Figures 3.2d, 6.2, 6.11): and when this obtains, then we will say – taking our lead from the diagram – that the map h *has a horseshoe*.

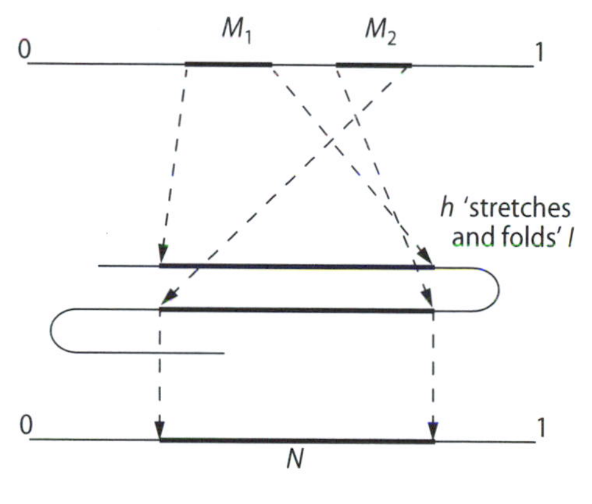

Figure 10.3 Stretching and folding producing a 'horseshoe'

Having a horseshoe in this technical sense requires a rather largish amount of stretching (at least a doubling on a single iteration); so we can not say, straight off, that f is chaotic just if it has a horseshoe. Still, a possible definition of chaos for a one-dimensional map f of the interval I is that f *eventually* has a horseshoe if iterated often enough – in other words, f is chaotic$_h$ just if f^k has a horseshoe for some k. It is not too hard to show that being chaotic$_h$ implies being chaotic$_d$ (Result H); and more, it gives us the exponential error inflation we think of as typical of true chaos. So some have adopted chaos$_h$ as their favoured definition, at least for the one-dimensional case.

(2) An apparently very different approach starts again from the thought that sensitive dependence is the crucial thing about chaos. But the fact that f satisfies (SDIC$_m$) doesn't tell us anything about how many orbits peel apart, or how fast: so we may well seek ways of quantifying this. Here's one way. Pretend that we can only observe the set I with a resolution ϵ (we can't distinguish points less than ϵ apart) – so pretend that, given a pair of orbits for f lasting for n iterations, we can only distinguish them if at some step their corresponding points are at least ϵ apart. Count the number of mutually distinguishable n-step orbits for f you can fit into I. If f inflates differences, then orbits starting from 'indistinguishable' points will become distinguishable. And the faster f inflates differences, the more orbits will be distinguishable within n steps. Further, if f really is chaotic, then we will expect the number of distinguishable orbits to keep on increasing as n increases – and indeed, we will expect this to be so, however small the level of resolution ϵ is set.

Call the limiting rate of increase in the number of distinguishable orbits as n increases the *topological entropy* of f – the greater the 'sensitivity' of the dynamics, the greater the entropy. Then we define a map to be chaotic$_{te}$ just if it has positive topological entropy. (See the following interlude for a little more explanation.)

(3) Finally, here is another, simpler and more direct, way of quantifying sensitive dependence for a one-dimensional map of the interval I. It could be that

$$(\mathrm{L}_m) \quad \text{if } x \approx y, \text{ then } |f^n(x) - f^n(y)| \approx |x - y|e^{\lambda n} \text{ where } \lambda > 0.$$

And then we will indeed have exponential error inflation, with the 'Liapunov exponent' λ giving the inflation rate. And since we are concentrating for the moment on maps of the interval into itself, orbits are necessarily confined – so that seems enough for chaos.

But intuitively, less will surely suffice (a map surely doesn't have to inflate at the same rate, and at every point, across the interval for it to have the sort of complexity we are interested in). So suppose e.g. we have

$$(\mathrm{L}'_m) \quad \text{if } x \approx y, \text{ then } |f^n(x) - f^n(y)| \approx |x - y|e^{\lambda(x)n} \quad \text{where } \lambda(x) \text{ is positive for almost all } x.$$

Then we will say that f is chaotic$_\lambda$. (For a remark on how to calculate Liapunov exponents, again see the following interlude.)

So we now have three more candidate definitions for chaos – at least for maps of the interval – with some complex, non-obvious, relations between them (and varying degrees of ease of application). For example, it is a deep and difficult result that, in the one-dimensional case, a function is chaotic$_{te}$ if and only if it is chaotic$_h$ – as Paul Glendinning puts it, 'In some sense, for maps of the interval … the horseshoe is not simply an example of chaotic behaviour, it is *the* example of chaotic behaviour.' (Glendinning 1994, 297).

Given its abstract generality and its role in generating powerful theorems, the notion of chaos$_{te}$ has some claim to centrality. On the other hand, determining topological entropies is usually no fun: Liapunov exponents tend to be much more tractable, more easily calculated. So a definition of chaos based on them may in practice be markedly more useful. Perhaps then we need both notions. Or indeed, perhaps we need all four of the concepts we have isolated so far …

A final mathematical interlude …

This last interlude sketches in a few more mathematical details.

We claimed (Result F) that Devaney's cut-down definition

(D_1) f is chaotic$_d$ if f has an invariant set K such that periodic points are dense in K, and f is topologically transitive on K

is equivalent to

(D_2) f is chaotic$_d$ if there is an invariant set K such that every pair of (non-empty) open subsets of K shares a periodic orbit.

It is trivial that (D_2) implies (D_1). For the reverse, note that if U and V are open sets, then – assuming transitivity – they share *an* orbit, i.e. there is a point $u \in U$ such that $f^k(u) \in V$. Now consider the (open) set of points W which get mapped by f^k into V, and consider the intersection of U and W. This is an open set, non-empty because it contains u, and – now assuming as well that periodic points are dense – it must contain a periodic point p. Which gives us a periodic point p in U that is mapped by f (after k steps) to V. So U and V share a periodic orbit. QED

We also claimed (Result G), that we couldn't delete 'periodic' from (D_2). Consider the map of the circle which takes a point with angular coordinate θ to the point $\theta + 2\pi\kappa$ where κ is irrational. Since κ is irrational, orbits never repeat, and any orbit eventually visits arbitrarily near any point on the circle. So, trivially, every pair of open subsets of the circle share *an* orbit. But there is no chaos, for there is no sensitive dependence. Orbits starting separated by δ always stay δ apart.

Result H, that chaos$_h$ implies chaos$_d$, is much more interesting. We need to show that (1) if a map eventually has a horseshoe then it has an invariant set, and (2) that the dynamics on this invariant set satisfies the Devaney conditions.

Suppose then that h has a horseshoe. That is, suppose there is an interval N which contains disjoint subintervals, M_0 and M_1, such that $h(M_0) = h(M_1) = N$. Consider the following construction (Figure 10.4).

Level 1. Put K_1 = the closure of $M_0 \cup M_1$.

Level 2. Put K_2 = the set of points in K_1 mapped by h back into K_1. In the general case, we will need to remove intervals from K_1 (as in the figure).

Level 3. Put K_3 = the set of points in K_2 mapped by h back into K_2 – constructed by removing intervals from K_2.

Keep on going. At each stage, we cut out intervals full of points whose orbits eventually leave K_1; K_n is the union of a lot of small remaining intervals (each one of which is mapped by h^n to N). While the limit set K_∞ – the intersection of all the K_i – is in general a fractal Cantor-like set, since we've removed intervals infinitely often.

K_∞ is invariant under h – for K_∞ is 'the set of points in K_∞ mapped by h back into K_∞'! And the dynamics on K_∞ is chaotic$_d$.

For suppose u is in K_∞ (and hence in all the K_i) and choose some small δ. Then for large enough n, there will be a K_n which has – as one of the little intervals that makes it up – an interval U which contains u and is no more than δ long. But $h^n(U) = N \supset U$. So, by the fixed point theorem (cf. Figure 6.6 and text), h^n has a fixed point in U, i.e. h has a period n point in U, call it v. Now, since the orbit of v never goes outside K_1, v also belongs to K_∞. So, in sum, for any u in K_∞, there is a periodic point v in K_∞ within δ – i.e. the periodic points are dense.

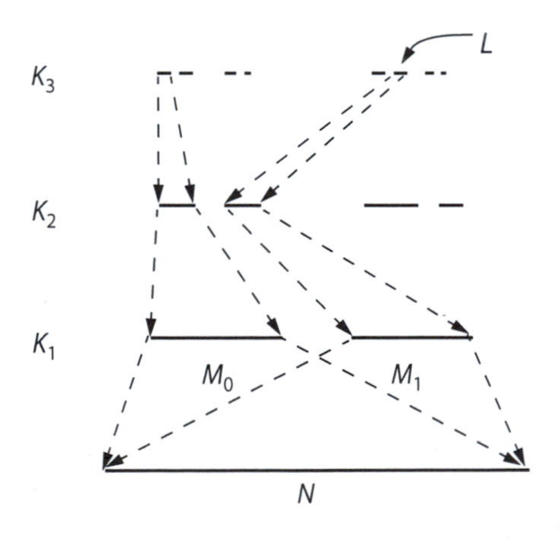

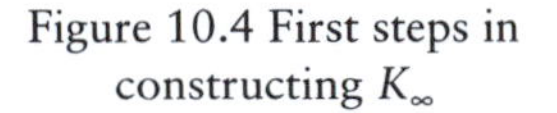
Figure 10.4 First steps in constructing K_∞

Transitivity can be proved similarly. So, if h has a horseshoe, it is chaotic$_d$. Which isn't yet quite what we wanted, which is to show that if a function *eventually* has a horseshoe, it is chaotic$_d$. But finishing the proof involves no interesting new ideas (let's leave it as an exercise!). Instead, we will comment on two features of the main line of the proof so far.

First, we informally noted long since (§3.3) how stretching and folding makes for fractal invariants: we here at last have a more formal indication of the point.

Second, note how we can identify the portions of K_3, say, as follows. L, in the figure, is the (largest) subset of M_1 which is first mapped into M_0 and then back into M_1. So we could index L by the label sequence '101'. Similarly, parts of K_5 could be labelled '00101', '11011' and the like. And likewise, in the limit, points in K_∞ can get infinite binary labels – as it might be '0010110 ...' which signifies a point which is mapped by h first into M_0, then into M_0, then into M_1 and so on. And the dynamics of h operates like a symbol-shift dynamics on these labels – it evidently maps the point '0010110 ...' to the point '010110 ...' and so on. (Hence, as announced before, the relevance of some kind of symbol-shift dynamics is not a special quirk of the logistic map.)

Finally, a little more explanation of the notion of topological entropy, and a remark about Liapunov exponents.

Suppose f is a continuous function from S into S (and now we are back to a more general case: S doesn't have to be a one-dimensional interval). Let's say that a subset $E \subset S$ is (n, ϵ)-separated if for all distinct x, y in E, the orbits of x and y get separated by at least ϵ before the n-th iteration of f (i.e., there is a k, $0 \leqslant k < n$, such that $|f^k(x) - f^k(x)| > \epsilon$). Now, since f is continuous, so is f^k, and that means that round any x we can choose some small-enough neighbourhood in which points *don't* get separated by ϵ. Our (n, ϵ)-separated set E, then, contains no intervals – it is a scattering of discrete points. Count them!

There will no doubt be various (n, ϵ)-separated sets in S – i.e. various ways of marking S with a grid of points which yield orbits which are distinguishable with resolution ϵ. So look at the (n, ϵ)-separated sets to find the one(s) with the maximum number of points (the largest cardinality). Let this number be $C(f, \epsilon, n)$.

If f inflates small differences, then as n increases, we will expect $C(f, \epsilon, n)$ to increase by some multiplicative factor, and to keep on increasing. So we want to define a limiting growth-rate for $C(f, \epsilon, n)$, call it $h(f, \epsilon)$ – so that

$$C(f, \epsilon, n) \approx Ce^{nh(f, \epsilon)}$$

Taking logarithms (and allowing progress to the limit to be subject to fluctuations) this motivates the definition

$$h(f, \epsilon) = \limsup_{n \to \infty} \frac{1}{n} \log C(f, \epsilon, n)$$

We then officially define the topological entropy $h(f)$ as the limit of $h(f, \epsilon)$ as $\epsilon \to 0$. And a map is chaotic$_{te}$ just if it has positive topological entropy.

Not, then, a straightforward notion – but one that has its roots in core thoughts about chaos. And further, one that arguably delivers the 'right' verdict in some problem cases. For example, $g_o{}^*$ which we constructed in the previous interlude will count as chaotic$_{te}$ though unbounded. While the other unbounded dynamics we mentioned – i.e. $f(x) = 2x$ on the half-line – is distinguished by the fact that there are a countably infinite number of (n, ϵ)-separated orbits for any n (so we can either take $h(f, \epsilon)$, and hence $h(f)$ to be undefined, or take it to equal $+\infty$, and then require for chaos positive, but non-infinite, entropy).

As to Liapunov exponents, it is worth adding one last mathematical remark (simply added to make connection with other presentations). In the §6.4 interlude 'More on period doubling', we saw in effect that the Liapunov exponent of f at x_0 equals

$$\log \lim_{n\to\infty} (\Lambda(x_0).\Lambda(x_1).\Lambda(x_2).\ldots.\Lambda(x_{n-1}))^{1/n} = \lim_{n\to\infty} 1/n \log (\Lambda(x_0).\Lambda(x_1).\Lambda(x_2).\ldots.\Lambda(x_{n-1}))$$

where $\Lambda(u)$ is the slope of f at u, i.e. $|df(x)/dx|_u$. But now note that by the familiar chain rule for differentiation

$$|df^n(x)/dx|_{x_0} = |df(u)/du|_{u_0}.|df^{n-1}(x)/dx|_{x_0} \text{ where } u = f^{n-1}(x), u_0 = f^{n-1}(x_0)$$
$$= \Lambda(x_{n-1})|df^{n-1}(x)/dx|_{x_0}$$

Whence, repeating the argument,

$$|df^n(x)/dx|_{x_0} = \Lambda(x_{n-1})\Lambda(x_{n-2}) \ldots \Lambda(x_1).\Lambda(x_0).$$

So, we can describe the Liapunov exponent with the standard equation (allowing now for some fluctuations as the limit is taken):

$$\lambda(x) = \limsup_{n\to\infty} \frac{1}{n} \log |df^n(x)/dx|_x$$

For reasonable mapping functions, this should be fairly readily estimable by standard computation techniques.

10.5 We have muddied the waters enough! We set out to define chaos, and we have found that – even if we just concentrate on one-dimensional maps of the interval – the task is far from straightforward. The standard Devaney definition has its shortcomings as a canonical account, and we already have thrown up three alternative ideas – and we haven't yet sorted out how best to extend all the candidate definitions to apply to discrete maps more generally. For instance, what is the generalization of the idea of 'having a horseshoe'? And how do we best handle the point that the map $f(x) = 2x$ on $[0, \infty)$ has positive Liapunov exponent, since $|f^n(x) - f^n(y)| = |x - y|e^{n\log 2}$, but it isn't chaotic (for we suggested earlier that we might not want to rule this case out *just* by insisting on confinement for chaos)?

But we won't pursue matters further. The issues now plainly turn on questions of theoretical fruitfulness, and what we need here is lots more mathematics – not more philosophical commentary. Still, we have probably already said enough to give the flavour of some avenues for exploration, and to strongly reinforce the suggestion that the case of 'chaos' is indeed like the examples of 'measure' or 'dimension' – there promises to be a number of possible fruitful approaches to the refinement and regimentation of the concept, with no one 'best' definition.

The complexities multiply again when we return at last to the case of continuous dynamics governed by differential equations as opposed to discrete difference maps, and ask what makes for chaos *there*. Though by this stage the reader will surely be happy enough to take this daunting claim on trust, rather than wade through yet more mathematical details.

We can tie the discrete and continuous cases together, by saying that continuous motion is chaotic if it induces a chaotic discrete map on a surface of section. Or we can give various direct definitions for the continuous case (analogues for some of the discrete definitions, plus further possibilities). But if we do want to plump for a single, rough, working criterion of chaos in the continuous case, then perhaps as good an option as any is to work in terms of Liapunov exponents again, which measure the rate at which trajectories spread apart or are compressed. In the general case, since there may be spreading in one direction and compression in others (see again Figure 3.2a), we will want to quantify the dynamics using a spectrum of numbers. And extending to the continuous case the sort of considerations just touched on at the end of the last interlude, we can give equations for these various exponents, leading to techniques for estimating them in practice.

Hence, characterizing a continuous chaotic dynamics as a system where there is a bounded attractor (to give confinement) on which the dynamics has at least one positive Liapunov exponent (so there is spreading and sensitive dependence) at least gives us a reasonably practicable criterion.

But by now it will hardly need stressing that this shouldn't be mistaken for the 'correct' or 'best' definition. There isn't one, any more than in the discrete case. And indeed, the further we pursue the mathematical investigations, the less important the definitional issue seems to become. Sharp new concepts are forged – like topological entropy (which is just one, indeed, of a whole family of entropy concepts), and many others – and as such new concepts become the focus of further work in their own right, the question of which ones should count as possible regimentations of our preformal notion of chaos rather loses its interest. Perhaps, in the end, we do best just to think of the preformal notion as a very rough pointer into a rich domain of enquiry, still young and in flux, still being mapped out – and leave it at that.

Notes

1 *Chaos introduced*

Stewart 1989, Hall 1992 and Lorenz 1993 provide elementary, almost maths-free, introductions to chaos. But for the purposes of this book, we need rather more. Strogatz 1994 is a particularly accessible mathematics text at undergraduate level: Ott 1993 takes the story further. The two volumes of Jackson 1990 make a fine sourcebook for additional mathematical information. Cvitanović 1989 and Hao 1990 are collections of reprinted journal articles, including some with a more experimental slant.

For a clear account of how to derive the Lorenz equations, see Appendix C of Hilborn 1994.

2 *Fractal intricacy*

Peitgen, Jürgens and Saupe 1992 is an accessible classroom text; at a more sophisticated mathematical level, Barnsley 1988 is particularly elegant. See Viscek 1992 on fractal growth phenomena.

The philosophical line in this chapter is similar to that taken independently in Shenker 1994; she is, however, rather less charitable in her reading of Mandelbrot.

3 *Intricacy and simplicity*

For the pivotal idea of 'stretching and folding' see e.g. Strogatz 1994, ch. 12. For more on the Rössler system, see also Jackson 1990, §7.11, and Holden and Muhamad 1986.

4 *Predictions*

Sparrow 1982 is an extended investigation of the Lorenz system. On 'shadowing theorems' see e.g. Barnsley 1988, §4.7, and Jackson 1990, §4.11.

5 *Approximate Truth*

For more on deflationism about truth, see Horwich 1990; for 'filtered disquotationalism', see Jackson et al. 1994. For an introduction to

Popper's attempts at defining verisimilitude, and the troubles that beset it, see e.g. O'Hear 1980, 47–56, or Niiniluoto 1987, Ch. 5.

6 *Universality*

Strogatz 1994, Ch. 10, is a good introduction to the complexities of the logistic map; but Devaney 1989 remains the key textbook. Cvitanović 1989 has an excellent introduction on universality, and also contains a number of key papers both on the experimental results and on Feigenbaum's fundamental theoretical results.

§6.7 is sceptical about the need to look for deep explanations of the recent upsurge of interest in chaotic dynamics among mathematical scientists. Compare, however, Kellert 1993, Ch. 5.

7 *Explanation*

Notions of 'robustness', i.e. of various kinds of stability, are crucial in dynamical theory: see e.g. Glendinning 1994, Ch. 2 for some distinctions. For a thought-provoking philosophical note, see Tavakol 1991. On supervaluationism, see the Introduction to Keefe and Smith 1997, and Fine 1975.

8 *Worldy chaos*

For the chaotic water-wheel, see e.g. Strogatz 1994, §9.1. For the BZ reaction, and much more on chemical chaos, see Scott 1991. On the reconstruction of attractors by time-delay coordinates, see Jackson 1991, §8.13; Ott 1993, §3.8; and for a book-length treatment, Abarbanel 1996.

For more on worldly chaos, see the papers collected in Part 2 of Cvitanović 1989, and in Part 2, §10, of Hao 1990.

Smith and Jones 1986, Ch. 18, introduces the philosophy of mind issues touched on in the concluding interlude. For more on the possible relevance of dynamical systems theory to psychology, see Horgan and Tienson 1996.

9 *Randomness*

For more on chaos and randomness, see also Ford 1986. Knuth 1981, Ch. 3, is an extensive survey of definitions of randomness. For more on the KCS treatment of randomness/patternlessness for finite sequences, see the papers reprinted in Parts 1 and 2 of Chaitin 1990. Earman 1986 is a modern classic, a wonderfully rich introduction to issues about determinism and randomness.

10 *Defining chaos*

For more philosophical discussion, see Batterman 1993. Various mathematical characterizations of chaos are given in e.g. Jackson 1990, §4.5; Ott 1993, §4.4; Glendinning 1994, Ch. 11. For another definition, and more on the ubiquity of symbol-shift dynamics, see also Wiggins 1988, Ch. 2. Other books give yet other definitions.

References

Abarbanel, H. 1996. *Analysis of Observed Chaotic Data*. New York: Springer.

Banks, J., J. Brooks, et al. 1992. On Devaney's definition of chaos. *American Mathematical Monthly* 99: 332–34.

Barnsley, M 1988. *Fractals Everywhere*. Boston: Academic Press.

Batterman, R. 1993. Defining chaos. *Philosophy of Science* 60: 43–66.

Crutchfield, J., J. Farmer, et al. 1986. What is chaos? *Scientific American* 255: 46–57.

Cvitanović, P. 1989. *Universality in Chaos*. 2nd ed. Bristol: Adam Hilger.

Devaney, R. 1989. *An Introduction to Chaotic Dynamical Systems*. 2nd ed. Redwood City, CA: Addison-Wesley.

Earman, J. 1986. *A Primer On Determinism*. Dordrecht: Reidel.

Falconer, K. 1990. *Fractal Geometry: Mathematical Foundations and Applications*. Chichester: John Wiley.

Fine, K. 1975. Vagueness, truth and logic. *Synthese* 30: 265–300.

Ford, J. 1986. Chaos: solving the unsolvable, predicting the unpredictable! In *Chaotic Dynamics and Fractals*, edited by M. Barnsley and S. Demko. San Diego: Academic Press.

Ford, J. 1989. What is chaos, that we should be mindful of it? In *The New Physics*, edited by P. Davies. Cambridge: Cambridge University Press.

Glendinning, P. 1994. *Stability, Instability and Chaos*. Cambridge: Cambridge University Press.

Goldstein, H. 1959. *Classical Mechanics*. Reading, MA: Addison-Wesley.

Hall, N., ed. 1992. *The New Scientist Guide to Chaos*. Harmondsworth: Penguin.

Hao, B-L. 1990. *Chaos II*. Singapore: World Scientific.

Herzen, A. 1968. *My Past and Thoughts*. Translated by C. Garnett, revised by H. Higgens. London: Chatto and Windus.

Hilborn, R. 1994. *Chaos and Nonlinear Dynamics*. New York: Oxford University Press.

Hobbs, J. 1993. Ex post facto explanations. *Journal of Philosophy* 90: 117–36.

Holden, A., and M. Muhamad. 1986. A graphical zoo of strange and peculiar attractors. In *Chaos*, edited by A. Holden. Manchester: Manchester University Press.

Horgan, T., and J. Tienson. 1996. *Connectionism and the Philosophy of Psychology*. Cambridge, MA: The MIT Press.

Horwich, P. 1990. *Truth*. Oxford: Blackwell.

Hunt, G. 1987. Determinism, predictability and chaos. *Analysis* 47: 129–33.

Jackson, E. Attlee. 1990. *Perspectives of Nonlinear Dynamics*. Cambridge: Cambridge University Press.

Jackson, F., G. Oppy, and M. Smith. 1994. Minimalism and truth aptness. *Mind* 103: 287–302.

Keefe, R., and P. Smith, eds. 1997. *Vagueness: A Reader.* Cambridge, MA: The MIT Press.

Kellert, S. 1993. *In the Wake of Chaos*. Chicago: University of Chicago Press.

Knuth, D. 1981. *The Art of Computer Programming*. 2nd ed. Vol. 2: Seminumerical Algorithms. Reading, MA: Addison-Wesley.

Lakatos, I. 1976. *Proofs and Refutations*. Cambridge: Cambridge University Press.

Lewis, D. 1986. *On the Plurality of Worlds*. Oxford: Basil Blackwell.

Lorenz, E. 1963. Deterministic nonperiodic flow. *Journal of the Atmospheric Sciences* 20: 130–41.

Lorenz, E. 1993. *The Essence of Chaos*. London: UCL Press.

Mandelbrot, B. 1967. How long is the coast of Britain? *Science* 155: 636–38.

Mandelbrot, B. 1983. *The Fractal Geometry of Nature*. New York: Freeman.

Martin-Löf, P. 1966. The definition of a random sequence. *Information and Control* 9: 602–19.

Miller, D. 1994. *Critical Rationalism*. Chicago: Open Court.

Montague, R. 1974. Deterministic theories. In *Formal Philosophy: Selected Papers of Richard Montague*, edited by R. Thomason. New Haven: Yale University Press.

Morton, A. 1991. The inevitability of folk psychology. In *Mind and Common Sense*, edited by R. Bogdan. Cambridge: Cambridge University Press).

Niiniluoto, I. 1987. *Truthlikeness*. Dordrecht: Reidel.

O'Hear, A. 1980. *Karl Popper.* London: Routledge and Kegan Paul.

Oddie, G. 1986. *Likeness to Truth*. Dordrecht: Reidel.

Ott, E. 1993. *Chaos in Dynamical Systems*. Cambridge: Cambridge University Press.

Peitgen, H-O., H. Jürgens, and D. Saupe. 1992. *Fractals for the Classroom.* New York: Springer-Verlag.

Railton, P. 1978. A deductive-nomological model of probabilistic explanation. *Philosophy of Science* 45: 206–26.

Roux, J., R. Simoyi, and H. Swinney. 1983. Observation of a strange attractor. *Physica* 8D: 257–66.

Ruelle, D. 1991. *Chance and Chaos*. Princeton, NJ: Princeton University Press.

Scott, S. 1991. *Chemical Chaos*. Oxford: Clarendon Press.

Shenker, O. 1994. Fractal geometry is not the geometry of nature. *Studies in History and Philosophy of Science* 25: 967–81.

Skarda, C., and W. Freeman. 1987. How brains make chaos in order to make sense of the world. *Behavioral and Brain Sciences* 10: 161–95.

Smith, P., and O. Jones. 1986. *The Philosophy of Mind.* Cambridge: Cambridge University Press.

Sparrow, C. 1982. *The Lorenz Equations: Bifurcations, Chaos and Strange Attractors*. New York: Springer.

Stewart, I. 1989. *Does God Play Dice?* Oxford: Basil Blackwell.

Strogatz, S. 1994. *Nonlinear Dynamics and Chaos*. Reading MA: Addison-Wesley.

Suppes, P. 1960. A comparison of the meaning and uses of models in mathematics and the empirical sciences. *Synthese* 12: 287–301.

Tavakol, R. 1991. Fragility and deterministic modelling in the exact sciences. *British Journal for the Philosophy of Science* 42: 147–56.

van Fraassen, B. 1980. *The Scientific Image*. Oxford: Clarendon Press.

Viscek, T. 1992. *Fractal Growth Phenomena*. Singapore: World Scientific.

Wiggins, S. 1988. *Global Bifurcations and Chaos*. New York: Springer-Verlag.

Index